Holt Science & Technology
Short Courses

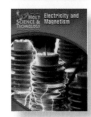

Teacher Edition WALK-THROUGH

Student Edition CONTENTS IN BRIEF

HOLT, RINEHART AND WINSTON

A Harcourt Education Company

Orlando • **Austin** • New York • San Diego • Toronto • London

Designed to meet the needs of all students

HOLT SCIENCE & TECHNOLOGY — Microorganisms, Fungi, and Plants

HOLT SCIENCE & TECHNOLOGY — Inside the Restless Earth

HOLT SCIENCE & TECHNOLOGY — Introduction to Matter

15 Short Courses

Holt Science & Technology: Short Course Series allows you to match your curriculum by choosing from 15 books covering life, earth, and physical sciences. The program reflects current curriculum developments and includes the strongest skills-development strand of any middle school science series. Students of all abilities will develop skills that they can use both in science as well as in other courses.

STUDENTS OF ALL ABILITIES RECEIVE THE READING HELP AND TAILORED INSTRUCTION THEY NEED.

- The *Student Edition* is accessible with a clean, easy-to-follow design and highlighted vocabulary words.
- Inclusion strategies and different learning styles help support all learners.
- Comprehensive **Section** and **Chapter Reviews** and **Standardized Test Preparation** allow students to practice their test-taking skills.
- **Reading Comprehension Guide** and **Guided Reading Audio CDs** help students better understand the content.

CROSS-DISCIPLINARY CONNECTIONS LET STUDENTS SEE HOW SCIENCE RELATES TO OTHER DISCIPLINES.

- **Mathematics, reading,** and **writing skills** are integrated throughout the program.
- Cross-discipline **Connection To** features show students how science relates to language arts, social studies, and other sciences.

A FLEXIBLE LABORATORY PROGRAM HELPS STUDENTS BUILD IMPORTANT INQUIRY AND CRITICAL-THINKING SKILLS.

- The laboratory program includes labs in each chapter, labs in the **LabBook** at the end of the text, six different lab books, and **Video Labs.**
- All labs are teacher-tested and rated by difficulty in the *Teacher Edition,* so you can be sure the labs will be appropriate for your students.
- A variety of labs, from **Inquiry Labs** to **Skills Practice Labs,** helps you meet the needs of your curriculum and work within the time constraints of your teaching schedule.

INTEGRATED TECHNOLOGY AND ONLINE RESOURCES EXPAND LEARNING BEYOND CLASSROOM WALLS.

- An **Enhanced Online Edition** or **CD-ROM Version** of the student text lightens your students' load.

- **SciLinks,** a Web service developed and maintained by the National Science Teachers Association (NSTA), contains current prescreened links directly related to the textbook.

- **Brain Food Video Quizzes** on videotape and DVD are game-show style quizzes that assess students' progress and motivate them to study.

- The **One-stop Planner**® CD-ROM with **Exam View**® **Test Generator** contains all of the resources you need including an *Interactive Teacher Edition,* worksheets, customizable lesson plans, **Holt Calendar Planner,** a powerful test generator, **Lab Materials QuickList Software,** and more.

- Spanish Resources include **Guided Reading Audio CD** in Spanish.

HOLT
CIENCIAS Y TECNOLOGÍA
LOS ANIMALES

EcoLabs

HOLT
SCIENCE & TECHNOLOGY

HOLT
CIENCIAS Y TECNOLOGÍA

Guided Reading Audio CD Program

Direct read of the student text

INTRODUCCIÓN A LA MATERIA
K

CHAPTER RESOURCE FILES FOR

Inside the Restless Earth

Skills Worksheets
- Directed Reading A
- Directed Reading B
- Vocabulary & Notes
- Section Reviews
- Chapter Review
- Reinforcement
- Critical Thinking

Assessments
- Section Quizzes
- Chapter Test A
- Chapter Test B
- Chapter Test C
- Performance-Based Assessment
- Standardized Test Preparation

Labs and Activities
- Datasheets for In-Text Labs
- Datasheets for Quick Labs
- Datasheets for LabBook
- Vocabulary Activity
- SciLinks® Activity

Teacher Resources
- Teacher Notes for Performance-Based Assessment
- Lab Notes and Answers
- Answer Keys
- Lesson Plans
- Test Item Listing for ExamView® Test Generator
- Teaching Transparencies
- Chapter Starter Transparencies
- Bellringer Transparencies
- Concept Mapping Transparencies

Life Science

| **A** MICROORGANISMS, FUNGI, AND PLANTS | **B** ANIMALS |

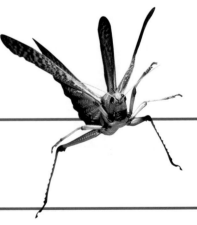

PROGRAM SCOPE AND SEQUENCE

Selecting the right books for your course is easy. Just review the topics presented in each book to determine the best match to your district curriculum.

C CELLS, HEREDITY, & CLASSIFICATION	**D** HUMAN BODY SYSTEMS & HEALTH	**E** ENVIRONMENTAL SCIENCE
Cells: The Basic Units of Life • Cells, tissues, and organs • Cell theory • Surface-to-volume ratio • Prokaryotic versus eukaryotic cells • Cell organelles	**Body Organization and Structure** • Homeostasis • Types of tissue • Organ systems • Structure and function of the skeletal system, muscular system, and integumentary system	**Interactions of Living Things** • Biotic versus abiotic parts of the environment • Producers, consumers, and decomposers • Food chains and food webs • Factors limiting population growth • Predator-prey relationships • Symbiosis and coevolution
The Cell in Action • Diffusion and osmosis • Passive versus active transport • Endocytosis versus exocytosis • Photosynthesis • Cellular respiration and fermentation • Cell cycle	**Circulation and Respiration** • Structure and function of the cardiovascular system, lymphatic system, and respiratory system • Respiratory disorders	**Cycles in Nature** • Water cycle • Carbon cycle • Nitrogen cycle • Ecological succession
Heredity • Dominant versus recessive traits • Genes and alleles • Genotype, phenotype, the Punnett square and probability • Meiosis • Determination of sex	**The Digestive and Urinary Systems** • Structure and function of the digestive system • Structure and function of the urinary system	**The Earth's Ecosystems** • Kinds of land and water biomes • Marine ecosystems • Freshwater ecosystems
Genes and Gene Technology • Structure of DNA • Protein synthesis • Mutations • Heredity disorders and genetic counseling	**Communication and Control** • Structure and function of the nervous system and endocrine system • The senses • Structure and function of the eye and ear	**Environmental Problems and Solutions** • Types of pollutants • Types of resources • Conservation practices • Species protection
The Evolution of Living Things • Adaptations and species • Evidence for evolution • Darwin's work and natural selection • Formation of new species	**Reproduction and Development** • Asexual versus sexual reproduction • Internal versus external fertilization • Structure and function of the human male and female reproductive systems • Fertilization, placental development, and embryo growth • Stages of human life	**Energy Resources** • Types of resources • Energy resources and pollution • Alternative energy resources
The History of Life on Earth • Geologic time scale and extinctions • Plate tectonics • Human evolution	**Body Defenses and Disease** • Types of diseases • Vaccines and immunity • Structure and function of the immune system • Autoimmune diseases, cancer, and AIDS	
Classification • Levels of classification • Cladistic diagrams • Dichotomous keys • Characteristics of the six kingdoms	**Staying Healthy** • Nutrition and reading food labels • Alcohol and drug effects on the body • Hygiene, exercise, and first aid	

Earth Science

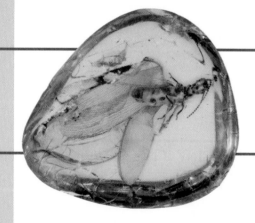

H WATER ON EARTH

The Flow of Fresh Water
- Water cycle
- River systems
- Stream erosion
- Life cycle of rivers
- Deposition
- Aquifers, springs, and wells
- Ground water
- Water treatment and pollution

Exploring the Oceans
- Properties and characteristics of the oceans
- Features of the ocean floor
- Ocean ecology
- Ocean resources and pollution

The Movement of Ocean Water
- Types of currents
- Characteristics of waves
- Types of ocean waves
- Tides

I WEATHER AND CLIMATE

The Atmosphere
- Structure of the atmosphere
- Air pressure
- Radiation, convection, and conduction
- Greenhouse effect and global warming
- Characteristics of winds
- Types of winds
- Air pollution

Understanding Weather
- Water cycle
- Humidity
- Types of clouds
- Types of precipitation
- Air masses and fronts
- Storms, tornadoes, and hurricanes
- Weather forecasting
- Weather maps

Climate
- Weather versus climate
- Seasons and latitude
- Prevailing winds
- Earth's biomes
- Earth's climate zones
- Ice ages
- Global warming
- Greenhouse effect

J ASTRONOMY

Studying Space
- Astronomy
- Keeping time
- Types of telescope
- Radioastronomy
- Mapping the stars
- Scales of the universe

Stars, Galaxies, and the Universe
- Composition of stars
- Classification of stars
- Star brightness, distance, and motions
- H-R diagram
- Life cycle of stars
- Types of galaxies
- Theories on the formation of the universe

Formation of the Solar System
- Birth of the solar system
- Structure of the sun
- Fusion
- Earth's structure and atmosphere
- Planetary motion
- Newton's Law of Universal Gravitation

A Family of Planets
- Properties and characteristics of the planets
- Properties and characteristics of moons
- Comets, asteroids, and meteoroids

Exploring Space
- Rocketry and artificial satellites
- Types of Earth orbit
- Space probes and space exploration

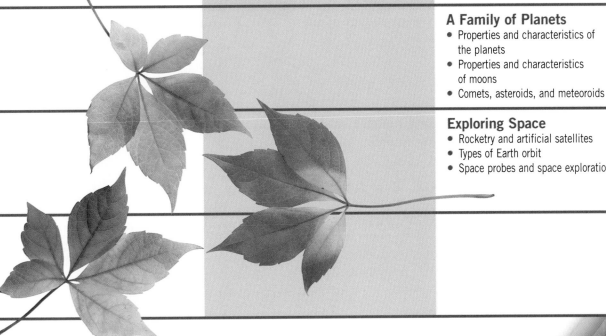

Physical Science

K INTRODUCTION TO MATTER	**L** INTERACTIONS OF MATTER
CHAPTER 1	
The Properties of Matter	**Chemical Bonding**
• Definition of matter	• Types of chemical bonds
• Mass and weight	• Valence electrons
• Physical and chemical properties	• Ions versus molecules
• Physical and chemical change	• Crystal lattice
• Density	
CHAPTER 2	
States of Matter	**Chemical Reactions**
• States of matter and their properties	• Writing chemical formulas and equations
• Boyle's and Charles's laws	• Law of conservation of mass
• Changes of state	• Types of reactions
	• Endothermic versus exothermic reactions
	• Law of conservation of energy
	• Activation energy
	• Catalysts and inhibitors
CHAPTER 3	
Elements, Compounds, and Mixtures	**Chemical Compounds**
• Elements and compounds	• Ionic versus covalent compounds
• Metals, nonmetals, and metalloids (semiconductors)	• Acids, bases, and salts
• Properties of mixtures	• pH
• Properties of solutions, suspensions, and colloids	• Organic compounds
	• Biomolecules
CHAPTER 4	
Introduction to Atoms	**Atomic Energy**
• Atomic theory	• Properties of radioactive substances
• Atomic model and structure	• Types of decay
• Isotopes	• Half-life
• Atomic mass and mass number	• Fission, fusion, and chain reactions
CHAPTER 5	
The Periodic Table	
• Structure of the periodic table	
• Periodic law	
• Properties of alkali metals, alkaline-earth metals, halogens, and noble gases	
CHAPTER 6	

M FORCES, MOTION, AND ENERGY

 N ELECTRICITY AND MAGNETISM

 O SOUND AND LIGHT

Matter in Motion
- Speed, velocity, and acceleration
- Measuring force
- Friction
- Mass versus weight

Introduction to Electricity
- Law of electric charges
- Conduction versus induction
- Static electricity
- Potential difference
- Cells, batteries, and photocells
- Thermocouples
- Voltage, current, and resistance
- Electric power
- Types of circuits

The Energy of Waves
- Properties of waves
- Types of waves
- Reflection and refraction
- Diffraction and interference
- Standing waves and resonance

Forces in Motion
- Terminal velocity and free fall
- Projectile motion
- Inertia
- Momentum

Electromagnetism
- Properties of magnets
- Magnetic force
- Electromagnetism
- Solenoids and electric motors
- Electromagnetic induction
- Generators and transformers

The Nature of Sound
- Properties of sound waves
- Structure of the human ear
- Pitch and the Doppler effect
- Infrasonic versus ultrasonic sound
- Sound reflection and echolocation
- Sound barrier
- Interference, resonance, diffraction, and standing waves
- Sound quality of instruments

Forces in Fluids
- Properties in fluids
- Atmospheric pressure
- Density
- Pascal's principle
- Buoyant force
- Archimedes' principle
- Bernoulli's principle

Electronic Technology
- Properties of semiconductors
- Integrated circuits
- Diodes and transistors
- Analog versus digital signals
- Microprocessors
- Features of computers

The Nature of Light
- Electromagnetic waves
- Electromagnetic spectrum
- Law of reflection
- Absorption and scattering
- Reflection and refraction
- Diffraction and interference

Work and Machines
- Measuring work
- Measuring power
- Types of machines
- Mechanical advantage
- Mechanical efficiency

Light and Our World
- Luminosity
- Types of lighting
- Types of mirrors and lenses
- Focal point
- Structure of the human eye
- Lasers and holograms

Energy and Energy Resources
- Forms of energy
- Energy conversions
- Law of conservation of energy
- Energy resources

Heat and Heat Technology
- Heat versus temperature
- Thermal expansion
- Absolute zero
- Conduction, convection, radiation
- Conductors versus insulators
- Specific heat capacity
- Changes of state
- Heat engines
- Thermal pollution

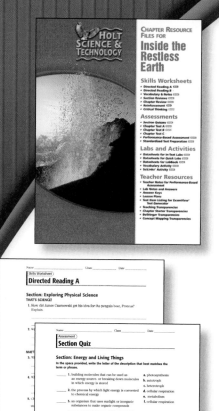

Program resources make teaching and learning easier.

CHAPTER RESOURCES

A *Chapter Resources book* accompanies each of the 15 *Short Courses*. Here you'll find everything you need to make sure your students are getting the most out of learning science—all in one book.

Skills Worksheets

- Directed Reading A: Basic
- Directed Reading B: Special Needs
- Vocabulary and Chapter Summary
- Section Reviews
- Chapter Reviews
- Reinforcement
- Critical Thinking

Labs & Activities

- Datasheets for Chapter Labs
- Datasheets for Quick Labs
- Datasheets for LabBook
- Vocabulary Activity
- SciLinks® Activity

Assessments

- Section Quizzes
- Chapter Tests A: General
- Chapter Tests B: Advanced
- Chapter Tests C: Special Needs
- Performance-Based Assessments
- Standardized Test Preparation

Teacher Resources

- Lab Notes and Answers
- Teacher Notes for Performance-Based Assessment
- Answer Keys
- Lesson Plans
- Test Item Listing for ExamView® Test Generator
- Full-color **Teaching Transparencies**, plus section **Bellringers, Concept Mapping,** and **Chapter Starter Transparencies.**

SPANISH RESOURCES

Spanish materials are available for each *Short Course:*

- *Student Edition*
- ***Spanish Resources*** booklet contains worksheets and assessments translated into Spanish with an English **Answer Key.**
- **Guided Reading Audio CD Program**

ONLINE RESOURCES

- *Enhanced Online Editions* engage students and assist teachers with a host of interactive features that are available anytime and anywhere you can connect to the Internet.
- **CNNStudentNews.com** provides award-winning news and information for both teachers and students.
- **SciLinks**—a Web service developed and maintained by the National Science Teachers Association—links you and your students to up-to-date online resources directly related to chapter topics.
- **go.hrw.com** links you and your students to online chapter activities and resources.
- **Current Science** articles relate to students' lives.

ADDITIONAL LAB AND SKILLS RESOURCES

- *Calculator-Based Labs* incorporates scientific instruments, offering students insight into modern scientific investigation.
- *EcoLabs & Field Activities* develops awareness of the natural world.
- *Holt Science Skills Workshop: Reading in the Content Area* contains exercises that target reading skills key.
- *Inquiry Labs* taps students' natural curiosity and creativity with a focus on the process of discovery.
- *Labs You Can Eat* safely incorporates edible items into the classroom.
- *Long-Term Projects & Research Ideas* extends and enriches lessons.
- *Math Skills for Science* provides additional explanations, examples, and math problems so students can develop their skills.
- *Science Skills Worksheets* helps your students hone important learning skills.
- *Whiz-Bang Demonstrations* gets your students' attention at the beginning of a lesson.

ADDITIONAL RESOURCES

- *Assessment Checklists & Rubrics* gives you guidelines for evaluating students' progress.
- *Holt Anthology of Science Fiction* sparks your students' imaginations with thought-provoking stories.
- *Holt Science Posters* visually reinforces scientific concepts and themes with seven colorful posters including **The Periodic Table of the Elements.**

- *Professional Reference for Teachers* contains professional articles that discuss a variety of topics, such as classroom management.
- *Program Introduction Resource File* explains the program and its features and provides several additional references, including lab safety, scoring rubrics, and more.
- *Science Fair Guide* gives teachers, students, and parents tips for planning and assisting in a science fair.
- *Science Puzzlers, Twisters & Teasers* activities challenge students to think about science concepts in different ways.

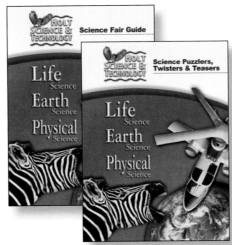

TECHNOLOGY RESOURCES

- *CNN Presents Science in the News: Video Library* helps students see the impact of science on their everyday lives with actual news video clips.
  - Multicultural Connections
  - Science, Technology & Society
  - Scientists in Action
  - Eye on the Environment
- *Guided Reading Audio CD Program*, available in English and Spanish, provides students with a direct read of each section.
- *HRW Earth Science Videotape* takes your students on a geology "field trip" with full-motion video.
- *Interactive Explorations CD-ROM Program* develops students' inquiry and decision-making skills as they investigate science phenomena in a virtual lab setting.

- *One-Stop Planner CD-ROM*® organizes everything you need on one disc, including printable worksheets, customizable lesson plans, a powerful test generator, **PowerPoint**® **Resources, Lab Materials QuickList Software, Holt Calendar Planner, Interactive Teacher Edition,** and more.
- *Science Tutor CD-ROMs* help students practice what they learn with immediate feedback.
- *Lab Videos* make it easier to integrate more experiments into your lessons without the preparation time and costs. Available on DVD and VHS.
- *Brain Food Video Quizzes* are game-show style quizzes that assess students' progress. Available on DVD and VHS.
- *Visual Concepts CD-ROMs* include graphics, animations, and movie clips that demonstrate key chapter concepts.

Science and Math Worksheets

The *Holt Science & Technology* program helps you meet the needs of a wide variety of students, regardless of their skill level. The following pages provide examples of the worksheets available to improve your students' science and math skills whether they already have a strong science and math background or are weak in these areas. Samples of assessment checklists and rubrics are also provided.

In addition to the skills worksheets represented here, *Holt Science & Technology* provides a variety of worksheets that are correlated directly with each chapter of the program. Representations of these worksheets are found at the beginning of each chapter in this *Teacher Edition*.

Many worksheets are also available on the Holt Web site. The address is **go.hrw.com**.

Science Skills Worksheets: Thinking Skills

BEING FLEXIBLE

USING YOUR SENSES
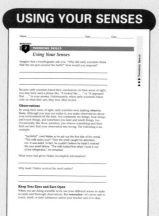

THINKING OBJECTIVELY

UNDERSTANDING BIAS

USING LOGIC
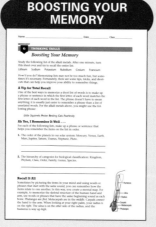

BOOSTING YOUR MEMORY

IMPROVING YOUR STUDY HABITS

READING A SCIENCE TEXTBOOK
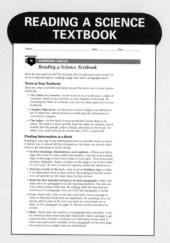

Science Skills Worksheets: Experimenting Skills

SAFETY RULES!

DOING A LAB WRITE-UP

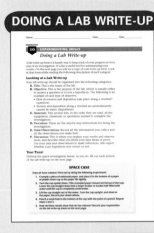

UNDERSTANDING VARIABLES

WORKING WITH HYPOTHESES

DESIGNING AN EXPERIMENT

USING THE INTERNATIONAL SYSTEM OF UNITS (SI)

MEASURING

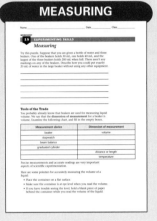

Science Skills Worksheets: Researching Skills

CHOOSING YOUR TOPIC

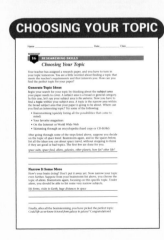

ORGANIZING YOUR RESEARCH

FINDING USEFUL SOURCES

RESEARCHING ON THE WEB

Science Skills Worksheets: Researching Skills (continued)

IDENTIFYING BIAS

TAKING NOTES

Science Skills Worksheets: Communicating Skills

SCIENCE WRITING

SCIENCE DRAWING

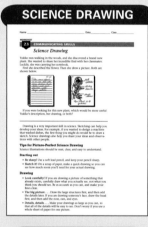

USING MODELS TO COMMUNICATE

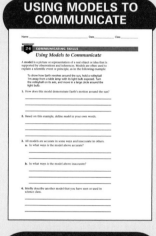

INTRODUCTION TO GRAPHS

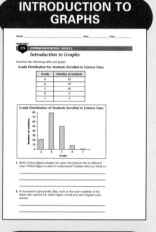

GRASPING GRAPHING

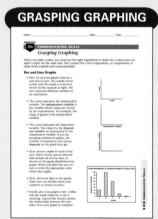

INTERPRETING YOUR DATA

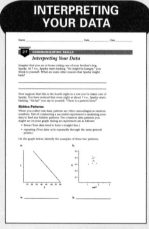

RECOGNIZING BIAS IN GRAPHS

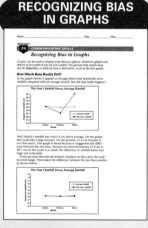

MAKING DATA MEANINGFUL

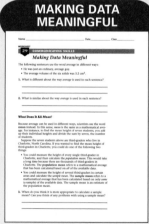

HINTS FOR ORAL PRESENTATIONS

Math Skills for Science

ADDITION AND SUBTRACTION

WORKSHEET 1 — MATH SKILLS

Addition Review

Addition is used to find the total of two or more quantities. The answer to an addition problem is known as the sum.

Add It Up!

Subtraction Review

WORKSHEET 2 — MATH SKILLS

Subtraction is used to take one number from another number. The answer to a subtraction problem is known as the difference.

Take It Away!

MULTIPLICATION

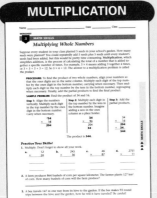

WORKSHEET 4 — MATH SKILLS

Multiplying Whole Numbers

Practice Your Skills!

WORKSHEET 3 — MATH SKILLS

A Shortcut for Multiplying Large Numbers

It's Your Turn!

Challenge Yourself!

DIVISION

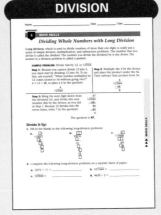

WORKSHEET 5 — MATH SKILLS

Dividing Whole Numbers with Long Division

Divide It Up!

WORKSHEET 6 — MATH SKILLS

Checking Division with Multiplication

Check It Out!

AVERAGES

WORKSHEET 7 — MATH SKILLS

What Is an Average?

Practice Your Skills!

WORKSHEET 8 — MATH SKILLS

Average, Mode, and Median

Get in the Mode!

POSITIVE AND NEGATIVE NUMBERS

WORKSHEET 9 — MATH SKILLS

Comparing Integers on a Number Line

Practice Your Skills!

WORKSHEET 12 — MATH SKILLS

Arithmetic with Positive and Negative Numbers

Part 1: Adding Positive and Negative Numbers

Add It Up!

FRACTIONS

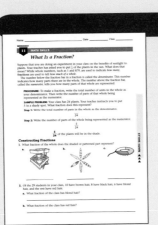

WORKSHEET 11 — MATH SKILLS

What Is a Fraction?

Constructing Fractions

WORKSHEET 13 — MATH SKILLS

Improper Fractions and Mixed Numbers

WORKSHEET 15 — MATH SKILLS

Multiplying and Dividing Fractions

Practice Your Skills!

WORKSHEET 10 — MATH SKILLS

Reducing Fractions to Lowest Terms

How Low Can You Go?

WORKSHEET 14 — MATH SKILLS

Adding and Subtracting Fractions

Part 1: Adding and Subtracting Fractions with the Same Denominator

Practice What You've Learned!

Part 2: Adding and Subtracting Fractions with Different Denominators

Math Skills for Science (continued)

RATIOS AND PROPORTIONS

What Is a Ratio?

Using Proportions and Cross-Multiplication

DECIMALS

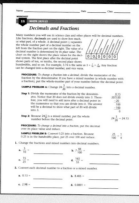

Decimals and Fractions

Arithmetic with Decimals

PERCENTAGES

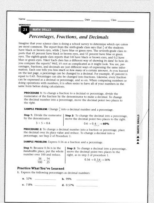

Parts of 100: Calculating Percentages

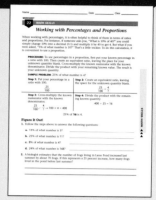

Working with Percentages and Proportions

Percentages, Fractions, and Decimals

POWERS OF 10

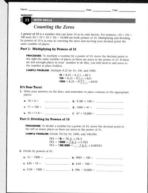

Counting the Zeros

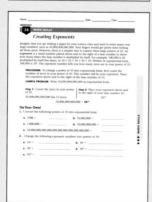

Creating Exponents

SCIENTIFIC NOTATION

What Is Scientific Notation?

Multiplying and Dividing in Scientific Notation

SI MEASUREMENT AND CONVERSION

What Is SI?

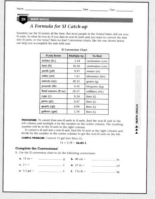

A Formula for SI Catch-up

Math Skills for Science (continued)

GEOMETRY

THE UNIT FACTOR AND DIMENSIONAL ANALYSIS

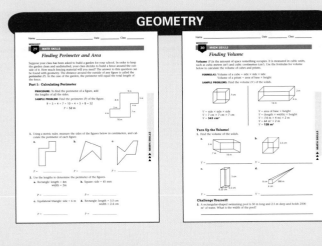

MATH IN SCIENCE: INTEGRATED SCIENCE

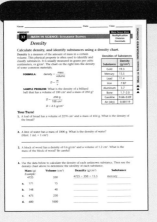

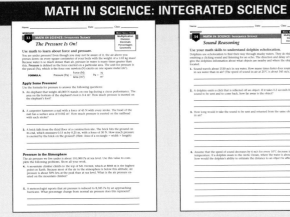

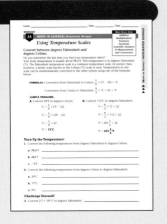

Math Skills for Science (continued)

MATH IN SCIENCE: LIFE SCIENCE

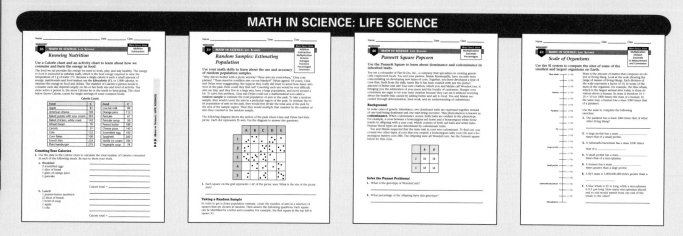

38 — *Knowing Nutrition*

39 — *Random Samples: Estimating Population*

40 — *Punnett Square Popcorn*

41 — *Scale of Organisms*

MATH IN SCIENCE: EARTH SCIENCE

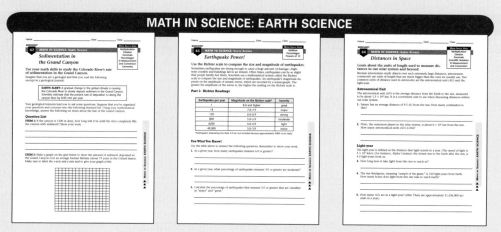

42 — *Sedimentation in the Grand Canyon*

43 — *Earthquake Power!*

44 — *Distances in Space*

45 — *Geologic Time Scale*

46 — *Mapping and Surveying*

Math Skills for Science (continued)

MATH IN SCIENCE: PHYSICAL SCIENCE

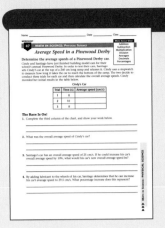
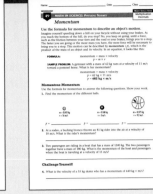
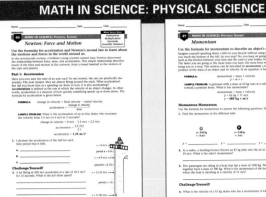

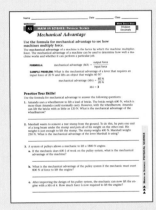

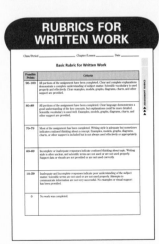
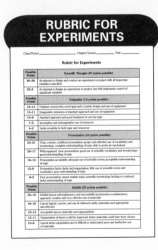

Assessment Checklist & Rubrics

The following is just a sample of over 50 checklists and rubrics contained in this booklet.

RUBRICS FOR WRITTEN WORK

RUBRIC FOR EXPERIMENTS

TEACHER EVALUATION OF COOPERATIVE LEARNING

TEACHER EVALUATION OF STUDENT PROGRESS

National Science Education Standards

The following lists show the chapter correlation of *Holt Science & Technology: Sound and Light* with the *National Science Education Standards* (grades 5–8).

Unifying Concepts and Processes	
Standard	**Chapter Correlation**
Systems, order, and organization Code: UCP 1	Chapter 2 2.2, 2.3
Evidence, models, and explanation Code: UCP 2	Chapter 3 3.1 Chapter 4 4.1, 4.2
Change, constancy, and measurement Code: UCP 3	Chapter 1 1.2, 1.3 Chapter 2 2.2 Chapter 3 3.1, 3.3
Form and function Code: UCP 5	Chapter 1 1.2 Chapter 2 2.4

Science as Inquiry	
Standard	**Chapter Correlation**
Abilities necessary to do scientific inquiry Code: SAI 1	Chapter 1 1.2 Chapter 2 2.1, 2.2, 2.3 Chapter 3 3.3
Understandings about scientific inquiry Code: SAI 2	Chapter 1 1.2 Chapter 2 2.2

Science in Personal and Social Perspectives

Standard	Chapter Correlation	
Personal health Code: SPSP 1	Chapter 2	2.1
	Chapter 3	3.2
Natural hazards Code: SPSP 3	Chapter 2	2.4
Science and technology in society Code: SPSP 5	Chapter 2	2.1, 2.2, 2.3
	Chapter 3	3.2
	Chapter 4	4.2, 4.3

History and Nature of Science

Standard	Chapter Correlation	
Science as human endeavor Code: HNS 1	Chapter 2	2.1, 2.2
History of science Code: HNS 3	Chapter 2	2.2

Physical Science Content Standards

Transfer of energy

Standard	Chapter Correlation	
Energy is a property of many substances and is associated with heat, light, electricity, mechanical motion, sound, nuclei, and the nature of a chemical. Energy is transferred in many ways. Code: PS 3a	**Chapter 1** **Chapter 2** **Chapter 3**	1.1, 1.2, 1.3 2.1, 2.2, 2.3 3.1, 3.2, 3.4
Light interacts with matter by transmission (including refraction), absorption, or scattering (including reflection). To see an object, light from that object—emitted or scattered from it—must enter the eye. Code: PS 3c	**Chapter 3** **Chapter 4**	3.3, 3.4 4.1, 4.2, 4.3
The sun is a major source of energy for changes on the earth's surface. The sun loses energy by emitting light. A tiny fraction of that light reaches the earth, transferring energy from the sun to the earth. The sun's energy arrives as light with a range of wavelengths, consisting of visible light, infrared, and ultraviolet radiation. Code: PS 3f	**Chapter 3**	3.1, 3.2

HOLT SCIENCE & TECHNOLOGY

Sound and Light

HOLT, RINEHART AND WINSTON

A Harcourt Education Company

Orlando • **Austin** • New York • San Diego • Toronto • London

Acknowledgments

Contributing Authors

Leila Dumas
Former Physics Teacher
Austin, Texas

Inclusion Specialist

Ellen McPeek Glisan
Special Needs Consultant
San Antonio, Texas

Safety Reviewer

Jack Gerlovich, Ph.D.
Associate Professor
School of Education
Drake University
Des Moines, Iowa

Academic Reviewers

Howard L. Brooks, Ph.D.
Professor of Physics & Astronomy
Department of Physics & Astronomy
DePauw University
Greencastle, Indiana

Simonetta Frittelli, Ph.D.
Associate Professor
Department of Physics
Duquesne University
Pittsburgh, Pennsylvania

David S. Hall, Ph.D.
Assistant Professor of Physics
Department of Physics
Amherst College
Amherst, Massachusetts

William H. Ingham, Ph.D.
Professor of Physics
James Madison University
Harrisonburg, Virginia

David Lamp, Ph.D.
Associate Professor of Physics
Physics Department
Texas Tech University
Lubbock, Texas

Mark Mattson, Ph.D.
Assistant Professor
Physics Department
James Madison University
Harrisonburg, Virginia

Richard F. Niedziela, Ph.D.
Assistant Professor of Chemistry
Department of Chemistry
DePaul University
Chicago, Illinois

H. Michael Sommermann, Ph.D.
Professor of Physics
Physics Department
Westmont College
Santa Barbara, California

Lab Testing

Barry L. Bishop
Science Teacher and Dept. Chair
San Rafael Junior High School
Ferron, Utah

Paul Boyle
Science Teacher
Perry Heights Middle School
Evansville, Indiana

Jennifer Ford
Science Teacher and Dept. Chair
North Ridge Middle School
North Richland Hills, Texas

Tracy Jahn
Science Teacher
Berkshire Junior-Senior High School
Canaan, New York

Edith C. McAlanis
Science Teacher and Dept. Chair
Socorro Middle School
El Paso, Texas

Kevin McCurdy, Ph.D.
Science Teacher
Elmwood Junior High School
Rogers, Arkansas

Terry J. Rakes
Science Teacher
Elmwood Junior High School
Rogers, Arkansas

Patricia McFarlane Soto
Science Teacher and Dept. Chair
G. W. Carver Middle School
Miami, Florida

Printed in the United States of America

ISBN 0-03-025587-2

4 5 6 7 048 08 07 06 05

0 Sound and Light

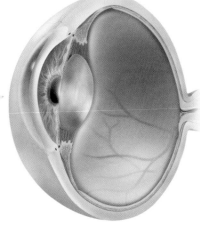

Labs and Activities

PRE-READING ACTIVITY

FOLDNOTES

Graphic Organizer

START-UP ACTIVITY

Quick Lab

Labs

INTERNET ACTIVITY

Go to go.hrw.com and type in the red keyword.

SCHOOL to HOME

READING STRATEGY

How to Use Your Textbook

Your Roadmap for Success with Holt Science and Technology

Reading Warm-Up

A Reading Warm-Up at the beginning of every section provides you with the section's objectives and key terms. The objectives tell you what you'll need to know after you finish reading the section.

Key terms are listed for each section. Learn the definitions of these terms because you will most likely be tested on them. Each key term is highlighted in the text and is defined at point of use and in the margin. You can also use the glossary to locate definitions quickly.

STUDY TIP Reread the objectives and the definitions to the key terms when studying for a test to be sure you know the material.

Get Organized

A Reading Strategy at the beginning of every section provides tips to help you organize and remember the information covered in the section. Keep a science notebook so that you are ready to take notes when your teacher reviews the material in class. Keep your assignments in this notebook so that you can review them when studying for the chapter test.

Be Resourceful—Use the Web

Interactions of Sound Waves

Have you ever heard of a sea canary? It's not a bird! It's a whale! Beluga whales are sometimes called sea canaries because of the many different sounds they make.

Dolphins, beluga whales, and many other animals that live in the sea use sound to communicate. Beluga whales also rely on reflected sound waves to find fish, crabs, and shrimp to eat. In this section, you will learn about reflection and other interactions of sound waves. You will also learn how bats, dolphins, and whales use sound to find food.

Reflection of Sound Waves

Reflection is the bouncing back of a wave after it strikes a barrier. You're probably already familiar with a reflected sound wave, otherwise known as an **echo.** The strength of a reflected sound wave depends on the reflecting surface. Sound waves reflect best off smooth, hard surfaces. Look at **Figure 1.** A shout in an empty gymnasium can produce an echo, but a shout in an auditorium usually does not.

The difference is that the walls of an auditorium are usually designed so that they absorb sound. If sound waves hit a flat, hard surface, they will reflect back. Reflection of sound waves doesn't matter much in a gymnasium. But you don't want to hear echoes while listening to a musical performance!

echo a reflected sound wave

Figure 1 Sound Reflection and Absorption

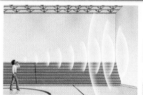

Sound waves easily reflect off the smooth, hard walls of a gymnasium. For this reason, you hear an echo.

In well-designed auditoriums, echoes are reduced by soft materials that absorb sound waves and by irregular shapes that scatter sound waves.

612 Chapter 21 The Nature of Sound

Internet Connect boxes in your textbook take you to resources that you can use for science projects, reports, and research papers. Go to scilinks.org, and type in the SciLinks code to get information on a topic.

Visit go.hrw.com Find worksheets, **Current Science**® magazine articles online, and other materials that go with your textbook at **go.hrw.com.** Click on the textbook icon and the table of contents to see all of the resources for each chapter.

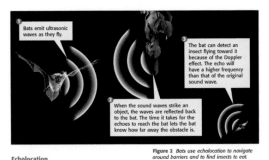

Figure 2 *Bats use echolocation to navigate around barriers and to find insects to eat.*

Use the Illustrations and Photos

Art shows complex ideas and processes. Learn to analyze the art so that you better understand the material you read in the text.

Tables and graphs display important information in an organized way to help you see relationships.

A picture is worth a thousand words. Look at the photographs to see relevant examples of science concepts that you are reading about.

Echolocation

Beluga whales use echoes to find food. The use of reflected sound waves to find objects is called **echolocation.** Other animals—such as dolphins, bats, and some kinds of birds—also use echolocation to hunt food and to find objects in their paths. **Figure 2** shows how echolocation works. Animals that use echolocation can tell how far away something is based on how long it takes sound waves to echo back to their ears. Some animals, such as bats, also make use of the Doppler effect to tell if another moving object, such as an insect, is moving toward it or away from it.

echolocation the process of using reflected sound waves to find objects; used by animals such as bats

✔ **Reading Check** How is echolocation useful to some animals? (See the Appendix for answers to Reading Checks.)

Echolocation

People use e using sonar (ranging). So **Figure 3** sho are used bec details about also help na can help oc

Resonance in Musical Instruments

Musical instruments use resonance to make sound. In wind instruments, vibrations are caused by blowing air into the mouthpiece. The vibrations make a sound, which is amplified when it forms a standing wave inside the instrument.

String instruments also resonate when they are played. An acoustic guitar, such as the one shown in **Figure 10,** has a hollow body. When the strings vibrate, sound waves enter the body of the guitar. Standing waves form inside the body of the guitar, and the sound is amplified.

Figure 10 *The body of a guitar resonates when the guitar is strummed.*

Answer the Section Reviews

Section Reviews test your knowledge of the main points of the section. Critical Thinking items challenge you to think about the material in greater depth and to find connections that you infer from the text.

STUDY TIP When you can't answer a question, reread the section. The answer is usually there.

Do Your Homework

Your teacher may assign worksheets to help you understand and remember the material in the chapter.

STUDY TIP Don't try to answer the questions without reading the text and reviewing your class notes. A little preparation up front will make your homework assignments a lot easier. Answering the items in the Chapter Review will help prepare you for the chapter test.

SECTION Review

Summary

● Echoes are reflected sound waves.

● Some animals can use echolocation to find food or to navigate around objects.

● People use echolocation technology in many underwater applications.

● Ultrasonography uses sound reflection for medical applications.

● Sound barriers and shock waves are created by interference.

● Standing waves form at an object's resonant frequencies.

● Resonance happens when a vibrating object causes a second object to vibrate at one of its resonant frequencies.

Using Key Terms

1. Use the following terms in the same sentence: *echo* and *echolocation.*

Complete each of the following sentences by choosing the correct term from the word bank.

| interference | standing wave |
| sonic boom | resonance |

2. When you pluck a string on a musical instrument, a(n) _____ forms.

3. When a vibrating object causes a nearby object to vibrate, _____ results.

Understanding Key Ideas

4. What causes an echo?
   a. reflection
   b. resonance
   c. constructive interference
   d. destructive interference

5. Describe a place in which you would expect to hear echoes.

6. How do bats use echoes to find insects to eat?

7. Give one example each of constructive and destructive interference of sound waves.

Math Skills

8. Sound travels through air at 343 m/s at 20°C. A bat emits an ultrasonic squeak and hears the echo 0.05 s later. How far away was the object that reflected it? (Hint: Remember that the sound must travel *to* the object and *back* to the bat.)

Critical Thinking

9. **Applying Concepts** Your friend is playing a song on a piano. Whenever your friend hits a certain key, the lamp on top of the piano rattles. Explain why the lamp rattles.

10. **Making Comparisons** Compare sonar and ultrasonography in locating objects.

SciLINKS. NSTA
Developed and maintained by the National Science Teachers Association

For a variety of links related to this chapter, go to www.scilinks.org

Topic: Interactions of Sound Waves
SciLinks code: HSM0804

617

SAFETY FIRST!

Exploring, inventing, and investigating are essential to the study of science. However, these activities can also be dangerous. To make sure that your experiments and explorations are safe, you must be aware of a variety of safety guidelines. You have probably heard of the saying, "It is better to be safe than sorry." This is particularly true in a science classroom where experiments and explorations are being performed. Being uninformed and careless can result in serious injuries. Don't take chances with your own safety or with anyone else's.

The following pages describe important guidelines for staying safe in the science classroom. Your teacher may also have safety guidelines and tips that are specific to your classroom and laboratory. Take the time to be safe.

Safety Rules!

Start Out Right

Always get your teacher's permission before attempting any laboratory exploration. Read the procedures carefully, and pay particular attention to safety information and caution statements. If you are unsure about what a safety symbol means, look it up or ask your teacher. You cannot be too careful when it comes to safety. If an accident does occur, inform your teacher immediately regardless of how minor you think the accident is.

If you are instructed to note the odor of a substance, wave the fumes toward your nose with your hand. Never put your nose close to the source.

Safety Symbols

All of the experiments and investigations in this book and their related worksheets include important safety symbols to alert you to particular safety concerns. Become familiar with these symbols so that when you see them, you will know what they mean and what to do. It is important that you read this entire safety section to learn about specific dangers in the laboratory.

Eye protection

Clothing protection

Hand safety

Heating safety

Electric safety

Chemical safety

Animal safety

Sharp object

Plant safety

x

Eye Safety

Wear safety goggles when working around chemicals, acids, bases, or any type of flame or heating device. Wear safety goggles any time there is even the slightest chance that harm could come to your eyes. If any substance gets into your eyes, notify your teacher immediately and flush your eyes with running water for at least 15 minutes. Treat any unknown chemical as if it were a dangerous chemical. Never look directly into the sun. Doing so could cause permanent blindness.

Avoid wearing contact lenses in a laboratory situation. Even if you are wearing safety goggles, chemicals can get between the contact lenses and your eyes. If your doctor requires that you wear contact lenses instead of glasses, wear eye-cup safety goggles in the lab.

Safety Equipment

Know the locations of the nearest fire alarms and any other safety equipment, such as fire blankets and eyewash fountains, as identified by your teacher, and know the procedures for using the equipment.

Neatness

Keep your work area free of all unnecessary books and papers. Tie back long hair, and secure loose sleeves or other loose articles of clothing, such as ties and bows. Remove dangling jewelry. Don't wear open-toed shoes or sandals in the laboratory. Never eat, drink, or apply cosmetics in a laboratory setting. Food, drink, and cosmetics can easily become contaminated with dangerous materials.

Certain hair products (such as aerosol hair spray) are flammable and should not be worn while working near an open flame. Avoid wearing hair spray or hair gel on lab days.

Sharp/Pointed Objects

Use knives and other sharp instruments with extreme care. Never cut objects while holding them in your hands. Place objects on a suitable work surface for cutting.

Be extra careful when using any glassware. When adding a heavy object to a graduated cylinder, tilt the cylinder so that the object slides slowly to the bottom.

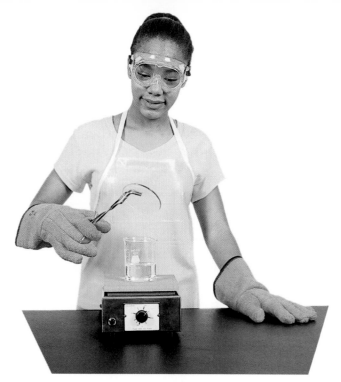

Heat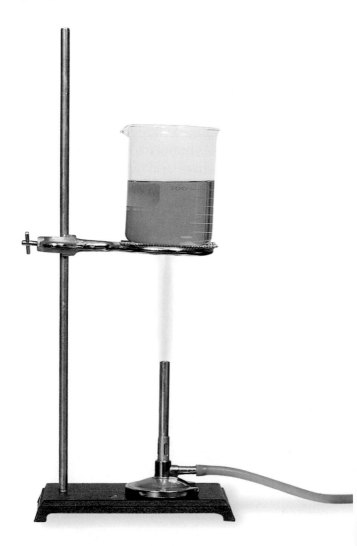

Wear safety goggles when using a heating device or a flame. Whenever possible, use an electric hot plate as a heat source instead of using an open flame. When heating materials in a test tube, always angle the test tube away from yourself and others. To avoid burns, wear heat-resistant gloves whenever instructed to do so.

Electricity

Be careful with electrical cords. When using a microscope with a lamp, do not place the cord where it could trip someone. Do not let cords hang over a table edge in a way that could cause equipment to fall if the cord is accidentally pulled. Do not use equipment with damaged cords. Be sure that your hands are dry and that the electrical equipment is in the "off" position before plugging it in. Turn off and unplug electrical equipment when you are finished.

Chemicals

Wear safety goggles when handling any potentially dangerous chemicals, acids, or bases. If a chemical is unknown, handle it as you would a dangerous chemical. Wear an apron and protective gloves when you work with acids or bases or whenever you are told to do so. If a spill gets on your skin or clothing, rinse it off immediately with water for at least 5 minutes while calling to your teacher.

Never mix chemicals unless your teacher tells you to do so. Never taste, touch, or smell chemicals unless you are specifically directed to do so. Before working with a flammable liquid or gas, check for the presence of any source of flame, spark, or heat.

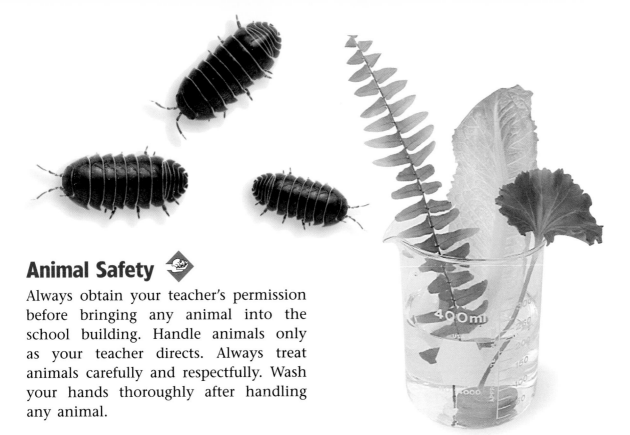

Animal Safety

Always obtain your teacher's permission before bringing any animal into the school building. Handle animals only as your teacher directs. Always treat animals carefully and respectfully. Wash your hands thoroughly after handling any animal.

Plant Safety

Do not eat any part of a plant or plant seed used in the laboratory. Wash your hands thoroughly after handling any part of a plant. When in nature, do not pick any wild plants unless your teacher instructs you to do so.

Glassware

Examine all glassware before use. Be sure that glassware is clean and free of chips and cracks. Report damaged glassware to your teacher. Glass containers used for heating should be made of heat-resistant glass.

The Energy of Waves
Chapter Planning Guide

Compression guide:
To shorten instruction because of time limitations, omit Section 3.

OBJECTIVES	LABS, DEMONSTRATIONS, AND ACTIVITIES	TECHNOLOGY RESOURCES
PACING • 90 min pp. 2–9 **Chapter Opener**	**SE Start-up Activity**, p. 3 (GENERAL)	**OSP Parent Letter** ■ (GENERAL) **CD Student Edition on CD-ROM** **CD Guided Reading Audio CD** ■ **TR Chapter Starter Transparency*** **VID Brain Food Video Quiz**
Section 1 The Nature of Waves • Describe how waves transfer energy without transferring matter. • Distinguish between waves that require a medium and waves that do not. • Explain the difference between transverse and longitudinal waves.	**TE Demonstration** Surfing Waves, p. 4 (GENERAL) **SE Connection to Astronomy** Light Speed, p. 6 (GENERAL) **SE Science in Action** Math, Social Studies, and Language Arts Activities, pp. 26–27 (GENERAL)	**CRF Lesson Plans*** **TR Bellringer Transparency*** **TR Wave Motion*** **TR Comparing Longitudinal and Transverse Waves*** **TR LINK TO EARTH SCIENCE** Primary Waves/Secondary Waves/Surface Waves* **CRF SciLinks Activity*** (GENERAL) **SE Internet Activity**, p. 7 (GENERAL)
PACING • 90 min pp. 10–13 **Section 2 Properties of Waves** • Identify and describe four wave properties. • Explain how frequency and wavelength are related to the speed of a wave.	**TE Demonstration** Pitch of a Rubber Band, p. 10 (GENERAL) **SE Quick Lab** Springy Waves, p. 11 (GENERAL) **CRF Datasheet for Quick Lab*** **TE Activity** Constructing a Transverse Wave, p. 11 (BASIC) **SE Skills Practice Lab** Wave Energy and Speed, p. 20 (GENERAL) **CRF Datasheet for Chapter Lab*** **SE Skills Practice Lab** Wave Speed, Frequency, and Wavelength, p. 124 (GENERAL) **CRF Datasheet for LabBook***	**CRF Lesson Plans*** **TR Bellringer Transparency*** **TR Measuring Wavelengths*** **TR Measuring Frequency*** **VID Lab Videos for Physical Science** **CD Science Tutor**
PACING • 45 min pp. 14–19 **Section 3 Wave Interactions** • Describe reflection, refraction, diffraction, and interference. • Compare destructive interference with constructive interference. • Describe resonance, and give examples.	**TE Activity** Wave Interactions, p. 14 (GENERAL) **TE Demonstration** Refraction, p. 15 (GENERAL) **SE School-to-Home Activity** What if Light Diffracted?, p. 16 (GENERAL) **TE Connection Activity** Real World, p. 16 (GENERAL) **TE Activity** AM and FM Radio Waves, p. 16 (ADVANCED) **TE Activity** Creating Interference, p. 17 (BASIC) **LB Whiz-Bang Demonstrations** Pitch Forks* (GENERAL) **LB Long-Term Projects & Research Ideas** A Whale of a Wave* (ADVANCED)	**CRF Lesson Plans*** **TR Bellringer Transparency*** **TR Constructive and Destructive Interference*** **CD Science Tutor**

PACING • 90 min

CHAPTER REVIEW, ASSESSMENT, AND STANDARDIZED TEST PREPARATION

CRF Vocabulary Activity* (GENERAL)
SE Chapter Review, pp. 22–23 (GENERAL)
CRF Chapter Review* ■ (GENERAL)
CRF Chapter Tests A* ■ (GENERAL), **B*** (ADVANCED), **C*** (SPECIAL NEEDS)
SE Standardized Test Preparation, pp. 24–25 (GENERAL)
CRF Standardized Test Preparation* (GENERAL)
CRF Performance-Based Assessment* (GENERAL)
OSP Test Generator (GENERAL)
CRF Test Item Listing* (GENERAL)

Online and Technology Resources

Visit **go.hrw.com** for a variety of free resources related to this textbook. Enter the keyword **HP5WAV**.

Holt Online Learning

Students can access interactive problem-solving help and active visual concept development with the *Holt Science and Technology* Online Edition available at **www.hrw.com**.

 Guided Reading Audio CD
Also in Spanish

A direct reading of each chapter for auditory learners, reluctant readers, and Spanish-speaking students.

 Science Tutor
CD-ROM

Excellent for remediation and test practice.

SKILLS DEVELOPMENT RESOURCES	SECTION REVIEW AND ASSESSMENT	CORRELATIONS
SE Pre-Reading Activity, p. 2 `GENERAL` **OSP** Science Puzzlers, Twisters & Teasers `GENERAL`		National Science Education Standards SAI 1, 2; PS 3a
CRF Directed Reading A* ■ `BASIC`, B* `SPECIAL NEEDS` **CRF** Vocabulary and Section Summary* ■ `GENERAL` **SE** Reading Strategy Discussion, p. 4 `GENERAL` **TE** Reading Strategy Prediction Guide, p. 5 `GENERAL` **TE** Inclusion Strategies, p. 7	**SE** Reading Checks, pp. 4, 6, 8 `GENERAL` **TE** Reteaching, p. 8 `BASIC` **TE** Quiz, p. 8 `GENERAL` **TE** Alternative Assessment, p. 8 `GENERAL` **SE** Section Review,* p. 9 ■ `GENERAL` **CRF** Section Quiz* ■ `GENERAL`	PS 3a
CRF Directed Reading A* ■ `BASIC`, B* `SPECIAL NEEDS` **CRF** Vocabulary and Section Summary* ■ `GENERAL` **SE** Reading Strategy Mnemonics, p. 10 `GENERAL` **TE** Reading Strategy Prediction Guide, p. 11 `GENERAL` **TE** Inclusion Strategies, p. 11 **SE** Math Focus Wave Calculations, p. 12 `GENERAL` **CRF** Reinforcement Worksheet Getting on the Same Frequency* `BASIC` **MS** Math Skills for Science Dividing Whole Numbers with Long Division* `GENERAL`	**SE** Reading Checks, pp. 11, 12 `GENERAL` **TE** Reteaching, p. 12 `BASIC` **TE** Quiz, p. 12 `GENERAL` **TE** Alternative Assessment, p. 12 `GENERAL` **SE** Section Review,* p. 13 ■ `GENERAL` **CRF** Section Quiz* ■ `GENERAL`	UCP 3, 5; SAI 1, 2; PS 3a; *Chapter Lab:* SAI 1, 2; PS 3a; *LabBook:* SAI 1, 2
CRF Directed Reading A* ■ `BASIC`, B* `SPECIAL NEEDS` **CRF** Vocabulary and Section Summary* ■ `GENERAL` **SE** Reading Strategy Reading Organizer, p. 14 `GENERAL` **CRF** Reinforcement Worksheet Makin' Waves* `BASIC` **CRF** Critical Thinking The Case of the Speeding Ticket* `ADVANCED`	**SE** Reading Checks, pp. 15, 17, 18 `GENERAL` **TE** Reteaching, p. 18 `BASIC` **TE** Quiz, p. 18 `GENERAL` **TE** Alternative Assessment, p. 18 `GENERAL` **SE** Section Review,* p. 19 ■ `GENERAL` **CRF** Section Quiz* ■ `GENERAL`	UCP 3; PS 3a

One-Stop Planner® CD-ROM

This CD-ROM package includes:
- Lab Materials QuickList Software
- Holt Calendar Planner
- Customizable Lesson Plans
- Printable Worksheets
- ExamView® Test Generator
- Interactive Teacher Edition
- Holt PuzzlePro® Resources
- Holt PowerPoint® Resources

SCILINKS.
NSTA

www.scilinks.org

Maintained by the **National Science Teachers Association.** See Chapter Enrichment pages for a complete list of topics.

Current Science®

Check out *Current Science* articles and activities by visiting the HRW Web site at **go.hrw.com.** Just type in the keyword **HP5CS20T.**

Classroom Videos

- **Lab Videos** demonstrate the chapter lab.
- **Brain Food Video Quizzes** help students review the chapter material.
- **CNN Videos** bring science into your students' daily life.

Visual Resources

CHAPTER STARTER TRANSPARENCY

This Really Happened!

On March 27, 1964, the most powerful earthquake ever recorded on the North American continent rocked Alaska. The quake started on land near the city of Anchorage, but seismic waves spread quickly in all directions, toppling buildings and ripping up roads.

The earthquake created a series of waves called tsunamis (tsoo NAH mees) in the Gulf of Alaska. In the deep water of the Gulf, the tsunamis were short and widely separated. But as these waves entered the shallow water surrounding Kodiak Island, off the coast of Alaska, they became taller and closer together. One of the tsunamis rose to a height of nearly 30 m! That's as tall as an eight-story building.

The powerful tsunamis pounded everything in their path. Eighty-one fishing boats were destroyed, and 86 others were damaged in the town of Kodiak. The destructive forces of the earthquake and tsunamis killed 21 people and caused $10 million in damage to Kodiak, making this the worst marine disaster in the town's 200-year history.

A tsunami is a dramatic example of the energy of waves. But waves affect your life in many common and less harmful ways. In fact, whenever you listen to music, talk with your friends, or watch a sunrise, you are experiencing the energy of waves. Read on to find out more about waves and how they affect your life every day.

BELLRINGER TRANSPARENCIES

Section: The Nature of Waves
What do you think of when you hear the word *wave?* Write a brief description in your **science journal** of what you think a wave is. Then write a short paragraph describing a time you might have experienced waves.

Section: Properties of Waves
Draw a longitudinal wave and a transverse wave in your **science journal**. Label the parts of each wave.

TEACHING TRANSPARENCIES

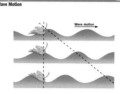

Wave Motion

Comparing Longitudinal and Transverse Waves

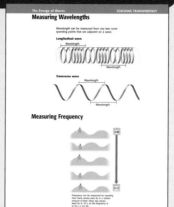

Measuring Wavelengths

Measuring Frequency

TEACHING TRANSPARENCIES

Primary Waves; Secondary Waves; Surface Waves

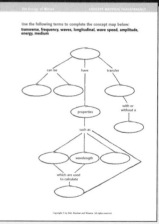

P waves move rock back and forth, which squeezes and stretches the rock, as they travel through the rock.

S waves shear rock side to side as they travel through the rock.

Surface waves move the ground much like ocean waves move water.

LINK TO EARTH SCIENCE

Chapter: Earthquakes

CONCEPT MAPPING TRANSPARENCY

Use the following terms to complete the concept map below:
transverse, frequency, waves, longitudinal, wave speed, amplitude, energy, medium

Planning Resources

LESSON PLANS

Lesson Plan SAMPLE

Section: Waves

Pacing
Regular Schedule: with lab(s):2 days without lab(s):1 days
Block Schedule: with lab(s): 1 1/2 days without lab(s):1 day

Objectives
1. Relate the seven properties of life to a living organism.
2. Describe seven themes that can help you to organize what you learn about biology.
3. Identify the tiny structures that make up all living organisms.
4. Differentiate between reproduction and heredity and between metabolism and homeostasis.

National Science Education Standards Covered
LSInter4:Cells have particular structures that underlie their functions.
LSMat1:Most cell functions involve chemical reactions.
LSBeh1:Cells store and use information to guide their functions.
UCP1:Cell functions are regulated.
SI1: Cells can differentiate and form complete multicellular organisms.
PS1:Species evolve over time.
ESS1: The great diversity of organisms is the result of more than 3.5 billion years of evolution.
ESS2: Natural selection and its evolutionary consequences provide a scientific explanation for the fossil record of ancient life forms as well as for the striking molecular similarities observed among the diverse species of living organisms.
ST1: The millions of different species of plants, animals, and microorganisms that live on Earth today are related by descent from common ancestors.
ST2: The energy for life primarily comes from the sun.
SPSP1: The complexity and organization of organisms accommodates the need for obtaining, transforming, transporting, releasing, and eliminating the matter and energy used to sustain the organism.
SPSP6: As matter and energy flows through different levels of organization of living systems—cells, organs, communities—and between living systems and the physical environment, chemical elements are recombined in different ways.
HNS1: Organisms have behavioral responses to internal changes and to external stimuli.

PARENT LETTER

SAMPLE

Dear Parent,

Your son's or daughter's science class will soon begin exploring the chapter entitled "The World of Physical Science." In this chapter, students will learn about how the scientific method applies to the world of physical science and the role of physical science in the world. By the end of the chapter, students should demonstrate a clear understanding of the chapter's main ideas and be able to discuss the following topics:

1. physical science is the study of energy and matter (Section 1)
2. the role of physical science in the world around them (Section 1)
3. careers that rely on physical science (Section 1)
4. the steps used in the scientific method (Section 2)
5. examples of technology (Section 2)
6. how the scientific method is used to answer questions and solve problems (Section 2)
7. how our knowledge of science changes over time (Section 2)
8. how models represent real objects or systems (Section 3)
9. examples of different ways models are used in science (Section 3)
10. the importance of the International System of Units (Section 4)
11. the appropriate units to use for particular measurements (Section 4)
12. how area and density are derived quantities (Section 4)

Questions to Ask Along the Way

You can help your son or daughter learn about these topics by asking interesting questions such as the following:

• What are some surprising careers that use physical science?
• What is a characteristic of a good hypothesis?
• When is it a good idea to use a model?
• Why do Americans measure things in terms of inches and yards and feet and meters?

ALSO IN SPANISH

TEST ITEM LISTING

TEST ITEM LISTING
The World of Science SAMPLE

MULTIPLE CHOICE

1. A limitation of models is that
   a. they are large enough to see.
   b. they do not act exactly like the things that they model.
   c. they are smaller than the things that they model.
   d. they model unfamiliar things.
   Answer: B Difficulty: 1 Section: 3 Objective: 2

2. The length 10 m is equal to
   a. 100 cm. c. 10,000 mm.
   b. 1,000 cm. d. Both (b) and (c)
   Answer: B Difficulty: 1 Section: 3 Objective: 2

3. To be valid, a hypothesis must be
   a. testable. c. made into a law.
   b. supported by evidence. d. Both (a) and (b)
   Answer: D Difficulty: 1 Section: 2 Objective: 2

4. The statement "Sheila has a stain on her shirt" is an example of a(n)
   a. law. c. observation.
   b. hypothesis. d. prediction.
   Answer: B Difficulty: 1 Section: 3 Objective: 2

5. A hypothesis is often developed out of
   a. observations. c. laws.
   b. experiments. d. Both (a) and (b)
   Answer: A Difficulty: 1 Section: 2 Objective: 2

6. How many milliliters are in 3.5 kL?
   a. 3,500 mL c. 3,500, 000 mL
   b. 0.0035 mL d. 35,000 mL
   Answer: B Difficulty: 1 Section: 3 Objective: 2

7. A map of Seattle is an example of a
   a. law. c. model.
   b. theory. d. unit.
   Answer: C Difficulty: 1 Section: 3 Objective: 2

8. A lab has the safety icons shown below. These icons mean that you should wear
   a. only safety goggles c. safety goggles and an apron.
   b. only a lab apron. d. safety goggles, a lab apron, and gloves.
   Answer: D Difficulty: 1 Section: 1 Objective: 2

9. The law of conservation of mass says the lot at mass before a chemical change is
   a. more than the total mass after the change.
   b. less than the total mass after the change.
   c. the same as the total mass after the change.
   d. not the same as the total mass after the change.
   Answer: C Difficulty: 1 Section: 2 Objective: 2

10. In which of the following areas might you find a geochemist at work?
    a. studying the chemistry of rocks c. studying fishes
    b. studying forestry d. studying the atmosphere
    Answer: A Difficulty: 1 Section: 2 Objective: 2

One-Stop Planner® CD-ROM

This CD-ROM includes all of the resources shown here and the following time-saving tools:

• *Lab Materials QuickList Software*
• *Customizable lesson plans*
• *Holt Calendar Planner*
• *The powerful ExamView® Test Generator*

Meeting Individual Needs

DIRECTED READING A
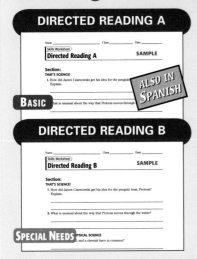
BASIC · ALSO IN SPANISH

DIRECTED READING B
SPECIAL NEEDS

VOCABULARY ACTIVITY
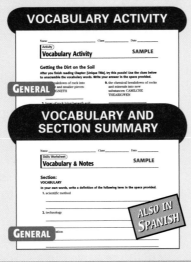
GENERAL

VOCABULARY AND SECTION SUMMARY
GENERAL · ALSO IN SPANISH

REINFORCEMENT

BASIC

CRITICAL THINKING
ADVANCED

SCILINKS ACTIVITY
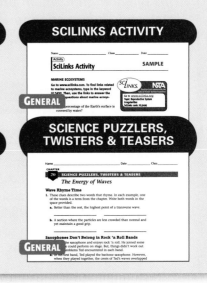
GENERAL

SCIENCE PUZZLERS, TWISTERS & TEASERS
GENERAL

Labs and Activities

LONG-TERM PROJECTS & RESEARCH IDEAS
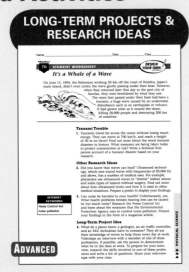
ADVANCED

WHIZ-BANG DEMONSTRATIONS

GENERAL

DATASHEETS FOR QUICK LABS

DATASHEETS FOR CHAPTER LABS

DATASHEETS FOR LABBOOK

Review and Assessments

SECTION QUIZ
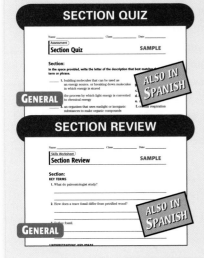
GENERAL · ALSO IN SPANISH

SECTION REVIEW
GENERAL · ALSO IN SPANISH

CHAPTER REVIEW
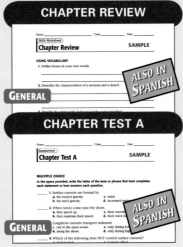
GENERAL · ALSO IN SPANISH

CHAPTER TEST A
GENERAL · ALSO IN SPANISH

CHAPTER TEST B
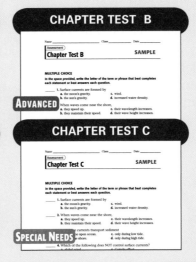
ADVANCED

CHAPTER TEST C
SPECIAL NEEDS

STANDARDIZED TEST PREPARATION
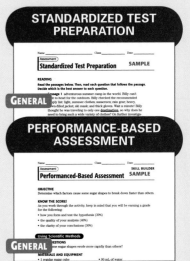
GENERAL

PERFORMANCE-BASED ASSESSMENT
GENERAL

This Chapter Enrichment provides relevant and interesting information to expand and enhance your presentation of the chapter material.

Section 1

The Nature of Waves

Tsunamis

- What many people call *tidal waves* are not caused by the tides. They are usually caused by undersea earthquakes, volcanic eruptions, landslides, or violent windstorms that occur over the open sea. A large wave produced by any of these phenomena is called a *tsunami*.

- Tsunamis often begin as a series of small waves no larger than 1 m high but traveling at great speed (188 m/s) over deep water. In deep water, a tsunami may not even be seen. However, when the depth of the water becomes less than the height of the wave, the wave builds into a wall of water that can reach a height of 30 m.

Mechanical Waves

- Some scientists call a wave "a wiggle in time and space." Mechanical waves are periodic disturbances that pass through matter. As a wave passes through matter, the material's particles vibrate about their rest positions, and only the energy moves through. Some of the wave's energy is used to do work on the particles. Mechanical waves eventually die out as their energy is dissipated.

Electromagnetic Waves

- An electromagnetic wave is a transverse wave comprising oscillating electric and magnetic fields at right angles to each other. Electromagnetic waves, like mechanical waves, are described in terms of wavelength and frequency. Types of electromagnetic waves (from longest wavelength to shortest) include radio waves, microwaves, infrared waves, visible light, ultra-violet light, X rays, and gamma rays.

Section 2

Properties of Waves

Waves

- A wave is produced by a vibrating object. For example, music is produced by vibrating strings or by vibrating columns of air. Electromagnetic waves can be produced by vibrating electrons.

- The frequency of the wave is equal to the frequency of the vibrating object. The square of the amplitude of the wave is proportional to the energy used to produce the wave.

- The speed of a wave depends on the medium through which the wave travels. For mechanical waves, the density, elasticity, and temperature of the medium affect the speed. A mechanical wave's speed increases with elasticity and decreases with density. For example, the speed of sound at 0°C is 331.5 m/s in air and 5,200 m/s in steel.

Is That a Fact!

- ◆ Heinrich Hertz (1857–1894) was a German physicist. The unit for describing frequency is named in his honor. Hertz's goal was to prove Maxwell's theory of electromagnetic radiation. While Hertz was performing his experiments, he discovered radio waves. This discovery led to the development of radio, radar, and television.

Longitudinal Waves and Wavelength

- Longitudinal waves do not have high and low points, but they do have regions of high pressure (compressions) and regions of low pressure (rarefactions).

- Sound waves are an example of longitudinal waves. When you strike a tuning fork, the prongs vibrate back and forth. Like a drum, the tuning fork sends out compressions when the prongs vibrate out and rarefactions when the prongs vibrate back. The compressions of the sound waves correspond to crests. Rarefactions of sound waves correspond to troughs. Therefore, the wavelength of a sound wave is the distance between adjacent compressions or adjacent expansions.

Is That a Fact!

◆ If you are listening to your favorite radio station, which has a frequency of 96.1 MHz (96,100,000 Hz), the electrons in the radio antenna are vibrating at the same frequency—96,100,000 vibrations/second.

Section 3

Wave Interactions
Christiaan Huygens

- Christiaan Huygens (1629–1695) was a Dutch mathematician, astronomer, and physicist. His life overlapped those of Galileo (1564–1642) and Newton (1642–1727). Many historians believe that Huygens's contributions to science were much broader than those of either Newton or Galileo.

- Huygens invented the pendulum clock and developed the first wave theory. Huygens's theory states that any point on a wave can be a source of a new disturbance.

Lasers

- In 1960, T. H. Maiman and others built the first successful laser. They used a large ruby crystal doped with chromium-ion impurities as the source of light. The electrons of the chromium ions were excited by a special flash lamp. The excited ions dropped to a lower energy level almost immediately and emitted photons of a single wavelength.

Is That a Fact!

◆ The word *refraction* comes from the Latin prefix *re-*, meaning "back," and *frangere*, meaning "to break."

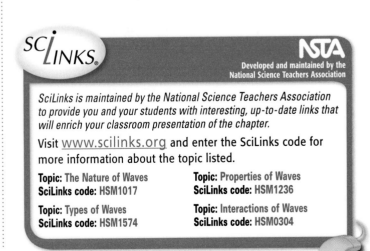

SciLinks is maintained by the National Science Teachers Association to provide you and your students with interesting, up-to-date links that will enrich your classroom presentation of the chapter.

Visit www.scilinks.org and enter the SciLinks code for more information about the topic listed.

Topic: The Nature of Waves
SciLinks code: HSM1017

Topic: Types of Waves
SciLinks code: HSM1574

Topic: Properties of Waves
SciLinks code: HSM1236

Topic: Interactions of Waves
SciLinks code: HSM0304

Overview

Tell students that this chapter will help them learn about the nature of waves. The chapter discusses the characteristics and properties of waves and the different kinds of wave interactions.

Assessing Prior Knowledge

Students should be familiar with the following topics:

- matter
- motion
- energy

Identifying Misconceptions

The motion of the medium (i.e., water) is frequently confused with the motion of the wave itself. Students also confuse the independent aspects of waves, such as amplitude, frequency, and velocity. A common belief is that a rapid oscillation ensures a large amplitude and fast velocity. Students may intuit that wave collisions will result in the permanent cancellation of both waves as if they were mechanical objects.

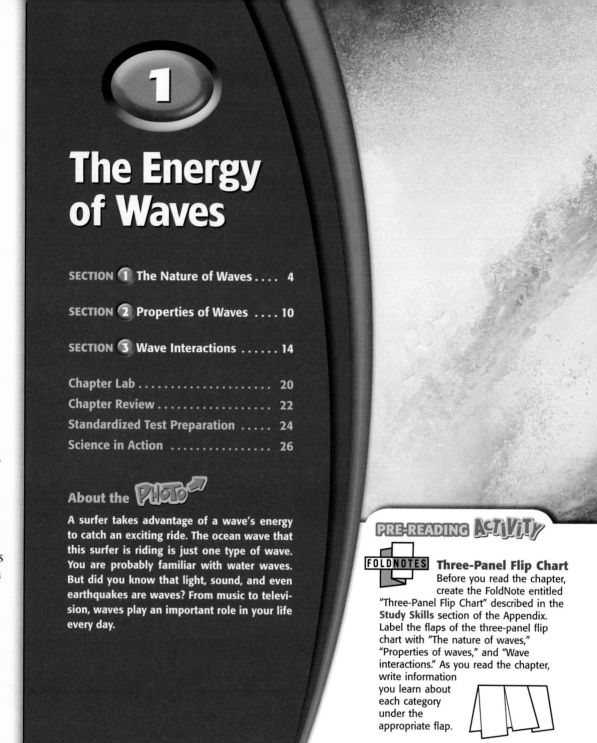

The Energy of Waves

About the PHOTO

A surfer takes advantage of a wave's energy to catch an exciting ride. The ocean wave that this surfer is riding is just one type of wave. You are probably familiar with water waves. But did you know that light, sound, and even earthquakes are waves? From music to television, waves play an important role in your life every day.

PRE-READING ACTIVITY

FOLDNOTES **Three-Panel Flip Chart**
Before you read the chapter, create the FoldNote entitled "Three-Panel Flip Chart" described in the **Study Skills** section of the Appendix. Label the flaps of the three-panel flip chart with "The nature of waves," "Properties of waves," and "Wave interactions." As you read the chapter, write information you learn about each category under the appropriate flap.

Standards Correlations

National Science Education Standards

The following codes indicate the National Science Education Standards that correlate to this chapter. The full text of the standards is at the front of the book.

Chapter Opener
SAI 1, 2; PS 3a

Section 1 The Nature of Waves
PS 3a

Section 2 Properties of Waves
UCP 3; SAI 1, 2; PS 3a; *LabBook:* SAI 1, 2

Section 3 Wave Interactions
UCP 3; PS 3a

Chapter Lab
SAI 1, 2; PS 3a

Chapter Review
PS 3a

Science in Action
ST 1, 2; HNS 3; PS 3a

START-UP ACTIVITY

MATERIALS

FOR EACH GROUP
- chair
- rope, 1–2 meters

Answers

1. The wave moves from one end of the rope to the other.

2. Each piece of rope moves up and down, that is, in a direction different from the wave. (If students have difficulty observing this, tie a piece of yarn to the rope, and have students watch only the yarn while waves are being made. The yarn will clearly move only up and down.)

3. The energy of the wave comes from shaking the rope. When students stop shaking the rope, the wave eventually stops moving.

START-UP ACTIVITY

Energetic Waves

In this activity, you will observe the movement of a wave. Then, you will determine the source of the wave's energy.

Procedure

1. Tie one end of a **piece of rope** to the back of a **chair.**

2. Hold the other end in one hand, and stand away from the chair so that the rope is almost straight but is not pulled tight.

3. Move the rope up and down quickly to create a wave. Repeat this step several times. Record your observations.

Analysis

1. In which direction does the wave move?

2. How does the movement of the rope compare with the movement of the wave?

3. Where does the energy of the wave come from?

The Energy of Waves   CHAPTER STARTER

This Really Happened!

On March 27, 1964, the most powerful earthquake ever recorded on the North American continent rocked Alaska. The quake started on land near the city of Anchorage, but the seismic waves spread quickly in all directions, toppling buildings and ripping up roads.

The earthquake created a series of waves called tsunamis (tsoo NAH mees) in the Gulf of Alaska. In the deep water of the Gulf, the tsunamis were short and widely separated. But as these waves entered the shallow water surrounding Kodiak Island, off the coast of Alaska, they became taller and closer together. One of the tsunamis rose to a height of nearly 30 m! That's as tall as an eight-story building.

The powerful tsunamis pounded everything in their path. Eighty-one fishing boats were destroyed, and 86 others were damaged in the town of Kodiak. The destructive forces of the earthquake and tsunamis killed 21 people and caused $10 million in damage to Kodiak, making this the worst marine disaster in the town's 200-year history.

A tsunami is a dramatic example of the energy of waves. But waves affect your life in many common and less harmful ways. In fact, whenever you listen to music, talk with your friends, or watch a sunrise, you are experiencing the energy of waves. Read on to find out more about waves and how they affect your life every day.

Chapter Starter Transparency
Use this transparency to help students begin thinking about waves and wave energy.

CHAPTER RESOURCES

Technology

 Transparencies
- Chapter Starter Transparency

READING SKILLS

 **Student Edition on CD-ROM**

 Guided Reading Audio CD
- English or Spanish

 **Classroom Videos**
- Brain Food Video Quiz

Workbooks

 Science Puzzlers, Twisters & Teasers
- The Energy of Waves GENERAL

Focus

Overview

In this section, students learn that waves are a means of transmitting energy. Students learn about different types of waves and their properties.

Bellringer

Have students answer the following question:

> What do you think of when you hear the word *wave*? Write a brief description of what you think a wave is. Then, write a short paragraph describing a time you might have experienced waves.

Discuss with students some of their answers.

Motivate

Demonstration— GENERAL

Surfing Waves Show students a video or photographs of people surfing the giant waves of California, Hawaii, South Africa, or Australia. Explain that this section introduces basic concepts about waves but that the waves that "break" as surfers ride them are not covered—scientists still do not fully understand why waves break. Studying waves around the world might make a fun career!
 Visual

READING WARM-UP

Objectives

- Describe how waves transfer energy without transferring matter.
- Distinguish between waves that require a medium and waves that do not.
- Explain the difference between transverse and longitudinal waves.

Terms to Learn

wave	transverse wave
medium	longitudinal wave

READING STRATEGY

Discussion Read this section silently. Write down questions that you have about this section. Discuss your questions in a small group.

wave a periodic disturbance in a solid, liquid, or gas as energy is transmitted through a medium

Figure 1 *Waves on a pond move toward the shore, but the water and the leaf floating on the surface only bob up and down.*

The Nature of Waves

Imagine that your family has just returned home from a day at the beach. You had fun playing in the ocean under a hot sun. You put some cold pizza in the microwave for dinner, and you turn on the radio. Just then, the phone rings. It's your friend calling to ask about homework.

In the events described above, how many different waves were present? Believe it or not, there were at least five! Can you name them? Here's a hint: A **wave** is any disturbance that transmits energy through matter or empty space. Okay, here are the answers: water waves in the ocean; light waves from the sun; microwaves inside the microwave oven; radio waves transmitted to the radio; and sound waves from the radio, telephone, and voices. Don't worry if you didn't get very many. You will be able to name them all after you read this section.

✓ *Reading Check* **What do all waves have in common?** (*See the Appendix for answers to Reading Checks.*)

Wave Energy

Energy can be carried away from its source by a wave. You can observe an example of a wave if you drop a rock in a pond. Waves from the rock's splash carry energy away from the splash. However, the material through which the wave travels does not move with the energy. Look at **Figure 1.** Can you move a leaf on a pond if you are standing on the shore? You can make the leaf bob up and down by making waves that carry enough energy through the water. But you would not make the leaf move in the same direction as the wave.

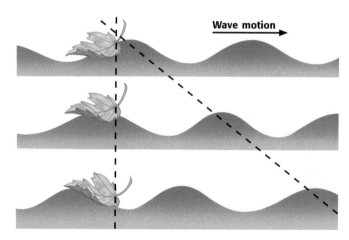

Wave motion

CHAPTER RESOURCES

Chapter Resource File

- Lesson Plan
- Directed Reading A **BASIC**
- Directed Reading B **SPECIAL NEEDS**

Technology

- Transparencies
  - Bellringer
  - Wave Motion
  - *LINK TO* **EARTH SCIENCE** Primary Waves/ Secondary Waves/Surface Waves

Answer to Reading Check
All waves are disturbances that transmit energy.

Waves and Work

As a wave travels, it does work on everything in its path. The waves in a pond do work on the water to make it move up and down. The waves also do work on anything floating on the water's surface. For example, boats and ducks bob up and down with waves. The fact that these objects move tells you that the waves are transferring energy.

Energy Transfer Through a Medium

Most waves transfer energy by the vibration of particles in a medium. A **medium** is a substance through which a wave can travel. A medium can be a solid, a liquid, or a gas. The plural of *medium* is *media*.

When a particle vibrates (moves back and forth, as in **Figure 2**), it can pass its energy to a particle next to it. The second particle will vibrate like the first particle does. In this way, energy is transmitted through a medium.

Sound waves need a medium. Sound energy travels by the vibration of particles in liquids, solids, and gases. If there are no particles to vibrate, no sound is possible. If you put an alarm clock inside a jar and remove all the air from the jar to create a vacuum, you will not be able to hear the alarm.

Other waves that need a medium include ocean waves, which move through water, and waves that are carried on guitar and cello strings when they vibrate. Waves that need a medium are called *mechanical waves*. **Figure 3** shows the effect of a mechanical wave in Earth's crust: an earthquake.

Figure 2 *A vibration is one complete back-and-forth motion of an object.*

medium a physical environment in which phenomena occur

Figure 3 *Earthquakes cause seismic waves to travel through Earth's crust. The energy they carry can be very destructive to anything on the ground.*

Is That a Fact!

Underwater Earthquakes Most earthquakes that take place on Earth happen underwater! Very large underwater earthquakes can cause huge waves called *tsunamis* that can be very destructive to anything on the shore.

MISCONCEPTION ALERT

Sounds in a Vacuum Students may believe that sounds can be heard in a vacuum. In many science-fiction movies, explosions and other sounds are heard in outer space. This is scientifically inaccurate.

READING STRATEGY — GENERAL

Prediction Guide Before students read this section, ask them whether the following statements are true or false:

• Light waves are mechanical waves because they must travel in a medium. (false)

• In space, no one can hear an explosion. (true)

• Water waves are a combination of longitudinal and transverse waves. (true)

LS Logical

Using the Figure — GENERAL

Discussion Encourage students to imagine that they are on the leaf shown in **Figure 1**. Ask them to describe their motions. Ask them how they could get to shore. (They would have to paddle.)

Would the waves carry them to shore? (no)

Teacher Note: Explain to students that in **Figure 1** the diagonal line shows the motion of a crest as it travels across a pond, and the vertical line shows that the leaf does not move with the wave. **LS** Visual

English Language Learners

CONNECTION to Earth Science — BASIC

Earthquakes Seismic waves carried through the Earth's crust occur when tectonic plates shift against each other suddenly. This phenomenon is known as an *earthquake*. Locations on Earth that are near *fault lines*, where tectonic plates meet, may be subject to small or large earthquakes.

CONNECTION to
Real World ——— GENERAL

Microwave Ovens Microwave ovens use electromagnetic waves to heat food. When the microwave energy penetrates a food item in a microwave oven, some of the energy from the waves causes water molecules to vibrate rapidly. The vibrating molecules are converting kinetic energy to thermal energy, which warms the rest of the particles in the food by conduction.

BRAIN FOOD

Electromagnetic Radiation
Albert Einstein, perhaps the world's most famous scientist, proposed that electromagnetic radiation could be viewed as a stream of particles, rather than as a wave of energy. He called these particles *photons*. In fact, Einstein proposed that energy is equivalent to mass and that photons effectively have mass. Experiments have proved that Einstein was right.

Figure 4 *Light waves are electromagnetic waves, which do not need a medium. Light waves from the Crab nebula, shown here, travel through the vacuum of space billions of miles to Earth, where they can be detected with a telescope.*

Energy Transfer Without a Medium

Some waves can transfer energy without going through a medium. Visible light is one example. Other examples include microwaves made by microwave ovens, TV and radio signals, and X rays used by dentists and doctors. These waves are *electromagnetic waves.*

Although electromagnetic waves do not need a medium, they can go through matter, such as air, water, and glass. The energy that reaches Earth from the sun comes through electromagnetic waves, which go through space. As shown in **Figure 4,** you can see light from stars because electromagnetic waves travel through space to Earth. Light is an electromagnetic wave that your eyes can see.

✓ Reading Check How do electromagnetic waves differ from mechanical waves?

CONNECTION TO
Astronomy

Light Speed Light waves from stars and galaxies travel great distances that are best expressed in light-years. A light-year is the distance a ray of light can travel in one year. Some of the light waves from these stars have traveled billions of light-years before reaching Earth. Do the following calculation in your **science journal:** If light travels at a speed of 300,000,000 m/s, what distance is a light-minute? (Hint: There are 60 s in a minute.) **ACTIVITY**

Answer to Reading Check
Electromagnetic waves do not require a medium.

Types of Waves

All waves transfer energy by repeated vibrations. However, waves can differ in many ways. Waves can be classified based on the direction in which the particles of the medium vibrate compared with the direction in which the waves move. The two main types of waves are *transverse waves* and *longitudinal* (LAHN juh TOOD'n uhl) *waves*. Sometimes, a transverse wave and a longitudinal wave can combine to form another kind of wave called a *surface wave*.

Transverse Waves

Waves in which the particles vibrate in an up-and-down motion are called **transverse waves.** *Transverse* means "moving across." The particles in this kind of wave move across, or perpendicularly to, the direction that the wave is going. To be *perpendicular* means to be "at right angles."

A wave moving on a rope is an example of a transverse wave. In **Figure 5,** you can see that the points along the rope vibrate perpendicularly to the direction the wave is going. The highest point of a transverse wave is called a *crest,* and the lowest point between each crest is called a *trough* (TRAWF). Although electromagnetic waves do not travel by vibrating particles in a medium, all electromagnetic waves are considered transverse waves. The reason is that the waves are made of vibrations that are perpendicular to the direction of motion.

INTERNET ACTIVITY

For another activity related to this chapter, go to **go.hrw.com** and type in the keyword **HP5WAVW.**

transverse wave a wave in which the particles of the medium move perpendicularly to the direction the wave is traveling

Figure 5 Motion of a Transverse Wave

A wave on a rope is a transverse wave because the particles of the medium vibrate perpendicularly to the direction the wave moves.

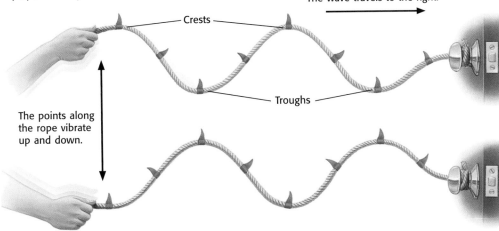

The wave travels to the right.

Crests

Troughs

The points along the rope vibrate up and down.

MISCONCEPTION
ALERT

Transverse Waves In this chapter, transverse waves are often represented as moving up and down. However, students should understand that not all transverse waves move in a vertical plane. A transverse wave is any wave in which the particles of the medium vibrate perpendicularly to the direction the wave moves.

Reteaching — **BASIC**

Wave Examples Have students brainstorm as many examples of waves as they can think of. For each example, have them state what makes it a wave and whether it is a transverse, longitudinal, or surface wave. **LS** Logical

Quiz — **GENERAL**

1. A wave is a disturbance that travels through _____ or _____. (space, matter)

2. A wave carries _____. (energy)

3. Waves that require a medium are called _____ waves. (mechanical)

4. Waves that do not require a medium are called _____ waves. (electromagnetic)

5. Waves produced by a combination of longitudinal and transverse waves are called _____ waves. (surface)

Alternative Assessment — **GENERAL**

Models of Waves Provide students with markers, construction paper, yarn, glue, and scissors. Have students use these materials to illustrate the following three wave types: longitudinal, transverse, and surface. Students should label their waves and the medium (if any) through which the wave is moving and should indicate compressions, rarefactions, crests, and troughs. **LS** Kinesthetic

---

Figure 6 **Comparing Longitudinal and Transverse Waves**

Pushing a spring back and forth creates a longitudinal wave, much the same way that shaking a rope up and down creates a transverse wave.

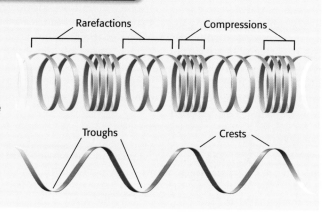

longitudinal wave a wave in which the particles of the medium vibrate parallel to the direction of wave motion

Longitudinal Waves

In a **longitudinal wave,** the particles of the medium vibrate back and forth along the path that the wave moves. You can make a longitudinal wave on a spring. When you push on the end of the spring, the coils of the spring crowd together. A part of a longitudinal wave where the particles are crowded together is called a *compression*. When you pull back on the end of the spring, the coils are pulled apart. A part where the particles are spread apart is a *rarefaction* (RER uh FAK shuhn). Compressions and rarefactions are like the crests and troughs of a transverse wave, as shown in **Figure 6.**

Sound Waves

A sound wave is an example of a longitudinal wave. Sound waves travel by compressions and rarefactions of air particles. **Figure 7** shows how a vibrating drum forms compressions and rarefactions in the air around it.

✔**Reading Check** What kind of wave is a sound wave?

Figure 7 *Sound energy is carried away from a drum by a longitudinal wave through the air.*

When the drumhead moves out after being hit, a compression is created in the air particles.

When the drumhead moves back in, a rarefaction is created.

CHAPTER RESOURCES

Technology

📦 **Transparencies**
• Comparing Longitudinal and Transverse Waves

Answer to Reading Check
A sound wave is a longitudinal wave.

Combinations of Waves

When waves form at or near the boundary between two media, a transverse wave and a longitudinal wave can combine to form a *surface wave*. An example is shown in **Figure 8.** Surface waves look like transverse waves, but the particles of the medium in a surface wave move in circles rather than up and down. The particles move forward at the crest of each wave and move backward at the trough.

Figure 8 *Ocean waves are surface waves. A floating bottle shows the circular motion of particles in a surface wave.*

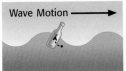

Wave Motion ⟶

SECTION Review

Summary

- A wave is a disturbance that transmits energy.
- The particles of a medium do not travel with the wave.
- Mechanical waves require a medium, but electromagnetic waves do not.
- Particles in a transverse wave vibrate perpendicularly to the direction the wave travels.
- Particles in a longitudinal wave vibrate parallel to the direction that the wave travels.

Using Key Terms

Complete each of the following sentences by choosing the correct term from the word bank.

transverse wave	wave
longitudinal wave	medium

1. In a ___, the particles vibrate parallel to the direction that the wave travels.

2. Mechanical waves require a ___ through which to travel.

3. Any ___ transmits energy through vibrations.

4. In a ___, the particles vibrate perpendicularly to the direction that the wave travels.

Understanding Key Ideas

5. Waves transfer
   a. matter.   c. particles.
   b. energy.   d. water.

6. Name a kind of wave that does not require a medium.

Critical Thinking

7. **Applying Concepts** Sometimes, people at a sports event do "the wave." Is this a real example of a wave? Why or why not?

8. **Making Inferences** Why can supernova explosions in space be seen but not heard on Earth?

Interpreting Graphics

9. Look at the figure below. Which part of the wave is the crest? Which part of the wave is the trough?

For a variety of links related to this chapter, go to www.scilinks.org

Topic: The Nature of Waves; Types of Waves
SciLinks code: HSM1017; HSM1574

CHAPTER RESOURCES

Chapter Resource File

- Section Quiz `GENERAL`
- Section Review `GENERAL`
- Vocabulary and Section Summary `GENERAL`
- SciLinks Activity `GENERAL`

SECTION 2

Focus

Overview

This section introduces properties common to all types of waves. Students explore frequency, wavelength, amplitude, energy content, and speed of waves.

 **Bellringer**

Before class, draw a longitudinal wave and a transverse wave on the board or the overhead projector. Have students draw the waves and label the parts of each one.

Motivate

Demonstration — GENERAL

Pitch of a Rubber Band Pluck a rubber band so that it makes a tone. Then, change the length of the rubber band so that the tone changes pitch. Ask students to explain what caused the pitch change. Have them observe the wave (the vibration) in the rubber band as you repeat the demonstration. Lead students to the conclusion that the tone changes as the frequency of the wave changes. Explain that the frequency is the number of waves produced in a given amount of time. (You can also use a guitar for this demonstration.) **Auditory**

SECTION 2

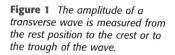 **READING WARM-UP**

Objectives
- Identify and describe four wave properties.
- Explain how frequency and wavelength are related to the speed of a wave.

Terms to Learn

amplitude   frequency
wavelength   wave speed

 READING STRATEGY

Mnemonics As you read this section, create a mnemonic device to help you remember the wave equation.

amplitude the maximum distance that the particles of a wave's medium vibrate from their rest position

Figure 1 The amplitude of a transverse wave is measured from the rest position to the crest or to the trough of the wave.

Properties of Waves

You are in a swimming pool, floating on your air mattress, enjoying a gentle breeze. Your friend does a "cannonball" from the high dive nearby. Suddenly, your mattress is rocking wildly on the waves generated by the huge splash.

The breeze generates waves in the water as well, but they are very different from the waves created by your diving friend. The waves made by the breeze are shallow and close together, while the waves from your friend's splash are tall and widely spaced. Properties of waves, such as the height of the waves and the distance between crests, are useful for comparing and describing waves.

Amplitude

If you tie one end of a rope to the back of a chair, you can create waves by moving the free end up and down. If you shake the rope a little, you will make a shallow wave. If you shake the rope hard, you will make a tall wave.

The **amplitude** of a wave is related to its height. A wave's amplitude is the maximum distance that the particles of a medium vibrate from their rest position. The rest position is the point where the particles of a medium stay when there are no disturbances. The larger the amplitude is, the taller the wave is. **Figure 1** shows how the amplitude of a transverse wave may be measured.

Larger Amplitude—More Energy

When using a rope to make waves, you have to work harder to create a wave with a large amplitude than to create one with a small amplitude. The reason is that it takes more energy to move the rope farther from its rest position. Therefore, a wave with a large amplitude carries more energy than a wave with a small amplitude does.

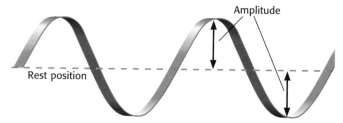

Amplitude

Rest position

CHAPTER RESOURCES

Chapter Resource File
- **Lesson Plan**
- **Directed Reading A** BASIC
- **Directed Reading B** SPECIAL NEEDS

Technology
- **Transparencies**
  - Bellringer
  - Measuring Wavelengths

MISCONCEPTION ALERT

Amplitude of a Longitudinal Wave The amplitude of a longitudinal wave is usually measured as a difference in pressure (air pressure or otherwise) between the compressions and rarefactions. Longitudinal waves are not discussed in terms of pressure in the text, so this method of measuring amplitude is not discussed.

Wavelength

Another property of waves is wavelength. A **wavelength** is the distance between any two crests or compressions next to each other in a wave. The distance between two troughs or rarefactions next to each other is also a wavelength. In fact, the wavelength can be measured from any point on a wave to the next corresponding point on the wave. Wavelength is measured the same way in both a longitudinal wave and a transverse wave, as shown in **Figure 2**.

Shorter Wavelength—More Energy

If you are making waves on either a spring or a rope, the rate at which you shake it will determine whether the wavelength is short or long. If you shake it rapidly back and forth, the wavelength will be shorter. If you are shaking it rapidly, you are putting more energy into it than if you were shaking it more slowly. So, a wave with a shorter wavelength carries more energy than a wave with a longer wavelength does.

✓ **Reading Check** How does shaking a rope at different rates affect the wavelength of the wave that moves through the rope? (*See the Appendix for answers to Reading Checks.*)

wavelength the distance from any point on a wave to an identical point on the next wave

---

wavelength the distance from any point on a wave to an identical point on the next wave

Springy Waves

1. Hold a coiled **spring toy** on the floor between you and a classmate so that the spring is straight. This is the rest position.

2. Move one end of the spring back and forth at a constant rate. Note the wavelength of the wave you create.

3. Increase the amplitude of the waves. What did you have to do? How did the change in amplitude affect the wavelength?

4. Now, shake the spring back and forth about twice as fast as you did before. What happens to the wavelength? Record your observations.

📖 **READING STRATEGY** — GENERAL

Prediction Guide Before students read this page, ask them, "Based on your experiences with water waves and from pictures of tsunamis, what do you think the height of a wave indicates?" Ask students to explain their predictions.

LS Logical/Intrapersonal

Answer to Reading Check

Shaking the rope faster makes the wavelength shorter; shaking the rope more slowly makes the wavelength longer.

ACTIVITY — BASIC

Constructing a Transverse Wave

Provide students with yarn and tape. Have them use the yarn to construct a transverse wave similar to the ones on these two pages. Ask them to increase the amplitude of the wave while keeping the frequency constant. (Students will need excess yarn for this step.) Have them explain what increasing the amplitude represents. Then, have them change the frequency, and ask them what happened to the wavelength when they changed the frequency. (The wavelength decreased with an increase in frequency, and it increased with a decrease in frequency.) **English Language Learners**

LS Kinesthetic

Figure 2 — Measuring Wavelengths

Wavelength can be measured from any two corresponding points that are adjacent on a wave.

Longitudinal wave

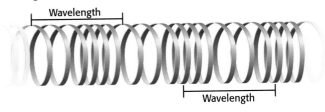

Wavelength
Wavelength

Transverse wave

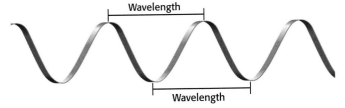

Wavelength
Wavelength

🔵 **INCLUSION Strategies**

• **Gifted and Talented**
• **Behavior Control Issues**

Point out that some students have a natural curiosity that drives them to want to know information beyond the textbook. Ask some of these students to research examples of jobs where a person would need to figure out amplitude, wavelength, frequency, or wave speed. **LS** Intrapersonal

---

MATERIAL

FOR EACH GROUP
• coiled spring toy or piece of rope

Answers

3. To increase the amplitude of the wave, the spring must be shaken farther with bigger motions. That is, the student must provide more energy to the wave. There should be no effect on wavelength when amplitude increases. (It may be difficult to increase amplitude without increasing frequency. If students increase frequency significantly, wavelength will change.)

4. The wavelength should become shorter as the frequency is increased.

Reteaching — BASIC

Measuring Wavelength and Amplitude Have students take turns coming to the board and alternating between drawing a wave and then labeling the wavelength and (if it is a transverse wave) amplitude of that wave. Have the students that draw the waves alternate between drawing a transverse and longitudinal wave. **LS** Visual/Interpersonal

Quiz — GENERAL

1. If wave speed is constant, as frequency increases, the _____ decreases. (wavelength)

2. _____ is the number of vibrations per second. (Frequency)

3. The distance between two corresponding points on consecutive waves is one _____. (wavelength)

4. As frequency increases, the _____ of the wave also increases. (energy)

Alternative Assessment — GENERAL

Concept Mapping Have students make a concept map explaining how amplitude, wavelength, and frequency are related to the energy of a wave. **LS** Visual

Answers to Math Focus

1. $f = v/\lambda = 12 \text{ cm/s} \div 3 \text{ cm} = 4/s$
= 4 Hz

2. $\lambda = v/f = 18 \text{ m/s} \div 5 \text{ Hz} = 18 \text{ m/s}$
$\div 5/s = 3.6 \text{ m}$

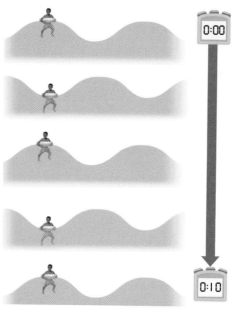

Figure 3 *Frequency can be measured by counting how many waves pass by in a certain amount of time. Here, two waves went by in 10 s, so the frequency is 2/10 s = 0.2 Hz.*

Frequency

Think about making rope waves again. The number of waves that you can make in 1 s depends on how quickly you move the rope. If you move the rope slowly, you make only a small number of waves each second. If you move it quickly, you make a large number of waves. The number of waves produced in a given amount of time is the **frequency** of the wave. Frequency is usually expressed in *hertz* (Hz). For waves, one hertz equals one wave per second (1 Hz = 1/s). **Figure 3** shows a wave with a frequency of 0.2 Hz.

✓ Reading Check If you make three rope waves per second, what is the frequency of the wave?

Higher Frequency—More Energy

To make high-frequency waves in a rope, you must shake the rope quickly back and forth. To shake a rope quickly takes more energy than to shake it slowly. Therefore, if the amplitudes are equal, high-frequency waves carry more energy than low-frequency waves.

Wave Speed

Wave speed is the speed at which a wave travels. Wave speed (v) can be calculated using wavelength (λ, the Greek letter *lambda*) and frequency (f), by using the *wave equation*, which is shown below:

$$v = \lambda \times f$$

MATH FOCUS

Wave Calculations Determine the wave speed of a wave that has a wavelength of 5 m and a frequency of 4 Hz.

Step 1: Write the equation for wave speed.

$$v = \lambda \times f$$

Step 2: Replace the λ and f with the values given in the problem, and solve.

$$v = 5 \text{ m} \times 4 \text{ Hz} = 20 \text{ m/s}$$

The equation for wave speed can also be rearranged to determine wavelength or frequency, as shown at top right.

$\lambda = \dfrac{v}{f}$ (Rearranged by dividing by f.)

$f = \dfrac{v}{\lambda}$ (Rearranged by dividing by λ.)

Now It's Your Turn

1. What is the frequency of a wave if the wave has a speed of 12 cm/s and a wavelength of 3 cm?

2. A wave has a frequency of 5 Hz and a wave speed of 18 m/s. What is its wavelength?

Is That a Fact!

Humans normally hear sounds with frequencies from 20 Hz to 20,000 Hz.

Answer to Reading Check

3 Hz

Frequency and Wavelength Relationship

Three of the basic properties of a wave are related to one another in the wave equation—wave speed, frequency, and wavelength. If you know any two of these properties of a wave, you can use the wave equation to find the third.

One of the things the wave equation tells you is the relationship between frequency and wavelength. If a wave is traveling a certain speed and you double its frequency, its wavelength will be cut in half. Or if you were to cut its frequency in half, the wavelength would be double what it was before. So, you can say that frequency and wavelength are *inversely* related. Think of a sound wave, traveling underwater at 1,440 m/s, given off by the sonar of a submarine like the one shown in **Figure 4**. If the sound wave has a frequency of 360 Hz, it will have a wavelength of 4.0 m. If the sound wave has twice that frequency, the wavelength will be 2.0 m, half as big.

The wave speed of a wave in a certain medium is the same no matter what the wavelength is. So, the wavelength and frequency of a wave depend on the wave speed, not the other way around.

Figure 4 *Submarines use sonar, sound waves in water, to locate underwater objects.*

frequency the number of waves produced in a given amount of time

wave speed the speed at which a wave travels through a medium

Answers to Section Review

1. Sample answer: amplitude: the maximum distance that the particles of a wave vibrate from their rest positions; frequency: the number of vibrations per second of a wave; wavelength: the distance between two corresponding points on a wave

2. a

3. Answers should include a diagram of a wave like the one in **Figure 1**, with amplitude indicated as shown in **Figure 1**, and wavelength indicated as shown in **Figure 2**.

4. $v = \lambda \times f = 2\text{ m} \times 6\text{ Hz} = 12\text{ m/s}$

5. Its wavelength must be short.

6. Both sounds will reach you at the same time: wave speed is not dependent on frequency, although the reverse is true.

SECTION Review

Summary

- Amplitude is the maximum distance the particles of a medium vibrate from their rest position.
- Wavelength is the distance between two adjacent corresponding parts of a wave.
- Frequency is the number of waves that pass a given point in a given amount of time.
- Wave speed can be calculated by multiplying the wave's wavelength by the frequency.

Using Key Terms

1. In your own words, write a definition for each of the following terms: *amplitude, frequency,* and *wavelength.*

Understanding Key Ideas

2. Which of the following results in more energy in a wave?
   - **a.** a smaller wavelength
   - **b.** a lower frequency
   - **c.** a shallower amplitude
   - **d.** a lower speed

3. Draw a transverse wave, and label how the amplitude and wavelength are measured.

Math Skills

4. What is the speed (*v*) of a wave that has a wavelength (λ) of 2 m and a frequency (*f*) of 6 Hz?

Critical Thinking

5. **Making Inferences** A wave has a low speed but a high frequency. What can you infer about its wavelength?

6. **Analyzing Processes** Two friends blow two whistles at the same time. The first whistle makes a sound whose frequency is twice that of the sound made by the other whistle. Which sound will reach you first?

SCLINKS.

Developed and maintained by the
National Science Teachers Association

For a variety of links related to this chapter, go to www.scilinks.org

Topic: Properties of Waves
SciLinks code: HSM1236

```
CHAPTER RESOURCES

Chapter Resource File

• Section Quiz  GENERAL
• Section Review  GENERAL
• Vocabulary and Section Summary  GENERAL
• Reinforcement Worksheet  BASIC
• Datasheet for Quick Lab
```

Focus

Overview

In this section, students explore how waves reflect, refract, or diffract when interacting with a different medium or barrier. Students also learn how waves interfere with other waves.

 **Bellringer**

Write v, f, and λ on the board. Have students write each symbol, what each symbol stands for, and how each symbol relates to the other two.

Motivate

ACTIVITY ———————— **GENERAL**

Wave Interactions Use two different types of springs. A long, tightly coiled spring, often called a snake, and a large, coiled spring toy are ideal. Using a strong cord, tie one end of each spring together, forming a "spring rope." Have two volunteers sit about 3 m apart on the floor. Have one student hold the spring rope on the floor while the other creates transverse waves by moving the spring from side to side. Ask the class to observe what happens at the boundary between the two springs when a wave passes from one spring to the other. (Students should see a change in the wave's speed and wavelength.) **LS Visual**

READING WARM-UP

Objectives

● Describe reflection, refraction, diffraction, and interference.

● Compare destructive interference with constructive interference.

● Describe resonance, and give examples.

Terms to Learn

reflection   interference
refraction   standing wave
diffraction   resonance

READING STRATEGY

Reading Organizer As you read this section, make a concept map by using the terms above.

---

reflection the bouncing back of a ray of light, sound, or heat when the ray hits a surface that it does not go through

Figure 1 These water waves are reflecting off the side of the container.

Wave Interactions

If you've ever seen a planet in the night sky, you may have had a hard time telling it apart from a star. Both planets and stars shine brightly, but the light waves that you see are from very different sources.

All stars, including the sun, produce light. But planets do not produce light. So, why do planets shine so brightly? The planets and the moon shine because light from the sun *reflects* off them. Without reflection, you would not be able to see the planets. Reflection is one of the wave interactions that you will learn about in this section.

Reflection

Reflection happens when a wave bounces back after hitting a barrier. All waves—including water, sound, and light waves—can be reflected. The reflection of water waves is shown in **Figure 1.** Light waves reflecting off an object allow you to see that object. For example, light waves from the sun are reflected when they strike the surface of the moon. These reflected waves allow us to enjoy moonlit nights. A reflected sound wave is called an *echo*.

Waves are not always reflected when they hit a barrier. If all light waves were reflected when they hit your eyeglasses, you would not be able to see anything! A wave is *transmitted* through a substance when it passes through the substance.

CHAPTER RESOURCES

Chapter Resource File

• Lesson Plan
• Directed Reading A **BASIC**
• Directed Reading B **SPECIAL NEEDS**

Technology

Transparencies
• Bellringer

Is That a Fact!

The reflection of sound is called an *echo*. If you are closer than 15 m to a reflecting wall, you cannot hear an echo. The brain requires a 0.1 s delay to perceive an echo, and at less than 15 m, the sound wave would be reflected back from the wall in less than 0.1 s.

Figure 2 *A light wave passing at an angle into a new medium—such as water—is refracted because the speed of the wave changes.*

Refraction

Try this simple activity: Place a pencil in a half-filled glass of water. Now, look at the pencil from the side. The pencil appears to be broken into two pieces! But as you can see when you take the pencil out of the water, it is still in one piece.

What you saw in this experiment was the result of the *refraction* of light waves. **Refraction** is the bending of a wave as the wave passes from one medium to another at an angle. Refraction of a flashlight beam as the beam passes from air to water is shown in **Figure 2.**

When a wave moves from one medium to another, the wave's speed changes. When a wave enters a new medium, the wave changes wavelength as well as speed. As a result, the wave bends and travels in a new direction.

✓ **Reading Check** What happens to a wave when it moves from one medium to another at an angle? (*See the Appendix for answers to Reading Checks.*)

Refraction of Different Colors

When light waves from the sun pass through a droplet of water in a cloud or through a prism, the light is refracted. But the different colors in sunlight are refracted by different amounts, so the light is *dispersed,* or spread out, into its separate colors. When sunlight is refracted this way through water droplets, you can see a rainbow. Why does that happen?

Although all light waves travel at the same speed through empty space, when light passes through a medium such as water or glass, the speed of the light wave depends on the wavelength of the light wave. Because the different colors of light have different wavelengths, their speeds are different, and they are refracted by different amounts. As a result, the colors are spread out, so you can see them individually.

refraction the bending of a wave as the wave passes between two substances in which the speed of the wave differs

---

CONNECTION TO Language Arts

WRITING SKILL **The Colors of the Rainbow** People have always been fascinated by the beautiful array of colors that results when sunlight strikes water droplets in the air to form a rainbow. The knowledge science gives us about how they form makes them no less breathtaking.

In the library, find a poem that you like about rainbows. In your **science journal,** copy the poem, and write a paragraph in which you discuss how your knowledge of refraction affects your opinion about the poem.

Teach

Demonstration — GENERAL

Refraction Display a large beaker half-filled with water. Place a ruler in the water, and rest it against the side. Have students observe the ruler from all sides. Have students record their observations and describe how the ruler appears to change when observed from different angles. Ask them to predict how the ruler would look if light entered the water perpendicularly to the surface. (The light would not refract, and the ruler would not look bent.) **LS** Logical

Answer to Reading Check
It refracts.

Using the Figure — GENERAL

Light Refraction Use **Figure 2** to show students how a wave bends during refraction. Point out that a wave that enters a new medium at a 90° angle from the surface of the medium will not refract because the entire wave enters the new medium at the same time. Therefore, the entire wave changes speed at the same time and does not bend. It may be helpful to draw a diagram similar to **Figure 2** that shows a wave passing from one medium to another at a 90° angle. **LS** Visual

---

MISCONCEPTION ///ALERT

Refraction and Frequency Students may think that as a wave enters a different medium and its speed changes, its frequency also changes. Be sure students understand that when a wave changes speed, its wavelength changes but its frequency remains the same. Frequency is dependent on the source, not on the medium.

Teach, *continued*

CONNECTION ACTIVITY
Real World ——————— GENERAL

Interfering Headphones In some occupations, workers are exposed to noise with amplitudes or frequencies that can be harmful. Now there are headphones available that create destructive interference to cancel the dangerous noise. These headphones receive the noise that the wearer hears, process the sounds, and then produce destructive interference to reduce or cancel the outside noise.

Have students think of and discuss situations in which these headphones would be useful.
LS Intrapersonal

ACTIVITY ——————— ADVANCED

AM and FM Radio Waves Have students research AM and FM radio waves. What are the differences between them? Why is it harder to receive FM broadcasts in some places? Why can AM broadcasts be heard for great distances under certain conditions? **LS** Logical

---

SCHOOL to HOME

What if Light Diffracted?
With an adult, take a walk around your neighborhood. Light waves diffract around corners of buildings much less than sound waves do. Imagine what would happen if light waves diffracted around corners much more than sound waves did. Write a paragraph in your **science journal** describing how this would change what you see and hear as you walk around your neighborhood.

ACTIVITY

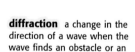

diffraction a change in the direction of a wave when the wave finds an obstacle or an edge, such as an opening

---

Figure 3 Diffraction Through an Opening

◀ If the barrier or opening is larger than the wavelength of the wave, there is only a small amount of diffraction.

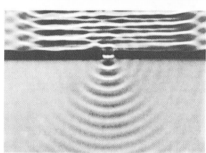

◀ If the barrier or opening is the same size or smaller than the wavelength of an approaching wave, the amount of diffraction is large.

Diffraction

Suppose you are walking down a city street and you hear music. The sound seems to be coming from around the corner, but you cannot see where the music is coming from because a building on the corner blocks your view. Why do sound waves travel around a corner better than light waves do?

Most of the time, waves travel in straight lines. For example, a beam of light from a flashlight is fairly straight. But in some circumstances, waves curve or bend when they reach the edge of an object. The bending of waves around a barrier or through an opening is known as **diffraction.**

If You Can Hear It, Why Can't You See It?

The amount of diffraction of a wave depends on its wavelength and the size of the barrier or opening the wave encounters, as shown in **Figure 3.** You can hear music around the corner of a building because sound waves have long wavelengths and are able to diffract around corners. However, you cannot see who is playing the music because the wavelengths of light waves are much shorter than sound waves, so light is not diffracted very much.

Answer to School-to-Home Activity
Sample answer: I could see around corners, so I could avoid running into people. I could see oncoming traffic. I could look under or around furniture or other objects to find missing items. But I wouldn't be able to hear people calling me from another room. I wouldn't be able to hear oncoming traffic.

Is That a Fact!

In 1962, New York City's Avery Fisher Hall opened, but there were problems with the hall's acoustics. There were too many echoes and dead spots caused by the interfering sound waves. In 1976, the hall's interior was removed and was replaced with an interior that produced better acoustics.

Interference

You know that all matter has volume. Therefore, objects cannot be in the same space at the same time. But waves are energy, not matter. So, more than one wave can be in the same place at the same time. In fact, two waves can meet, share the same space, and pass through each other! When two or more waves share the same space, they overlap. The result of two or more waves overlapping is called **interference. Figure 4** shows what happens when waves occupy the same space and interfere with each other.

interference the combination of two or more waves that results in a single wave

Constructive Interference

Constructive interference happens when the crests of one wave overlap the crests of another wave or waves. The troughs of the waves also overlap. When waves combine in this way, the energy carried by the waves is also able to combine. The result is a new wave that has higher crests and deeper troughs than the original waves had. In other words, the resulting wave has a larger amplitude than the original waves had.

✓ **Reading Check** How does constructive interference happen?

Figure 4 Constructive and Destructive Interference

Constructive Interference When waves combine by constructive interference, the combined wave has a larger amplitude.

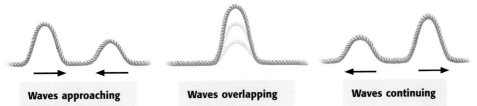

| Waves approaching | Waves overlapping | Waves continuing |

Destructive Interference When two waves with the same amplitude combine by destructive interference, they cancel each other out.

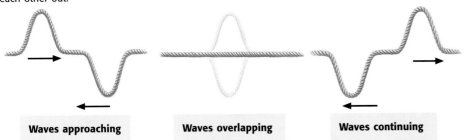

| Waves approaching | Waves overlapping | Waves continuing |

MISCONCEPTION ALERT

Interference of Waves Interference occurs only in the region where the waves overlap. After the waves pass through this region, they continue on as if they had never met.

Reteaching ── BASIC

Reteaching ── BASIC

Wave Interactions Table On the board, make a table with three columns and five rows. Fill in the first row with "Wave interaction," "Cause," and "Effect." Fill in the first column with "Reflection," "Refraction," "Diffraction," and "Interference." Call on different students to fill in the different spaces in the table, indicating what causes each wave interaction and what effect each has on a wave.
LS Visual

Quiz ── GENERAL

1. When a wave bounces back from a barrier, _____ has occurred. (reflection)

2. _____ occurs when a wave bends as it passes at an angle from one medium to a different medium. (Refraction)

3. _____ happens when two or more waves overlap. (Interference)

Alternative Assessment ── GENERAL

Concept Mapping Have students create a concept map using as many vocabulary words from this chapter as they can. Then, provide groups of students with yarn, poster board, glue, tape, and scissors. Each group should construct a model for the concepts on its map.
LS Visual

Destructive Interference

Destructive interference happens when the crests of one wave and the troughs of another wave overlap. The new wave has a smaller amplitude than the original waves had. When the waves involved in destructive interference have the same amplitude and meet each other at just the right time, the result is no wave at all.

Standing Waves

If you tie one end of a rope to the back of a chair and move the other end up and down, the waves you make go down the rope and are reflected back. If you move the rope at certain frequencies, the rope appears to vibrate in loops, as shown in **Figure 5.** The loops come from the interference between the wave you made and the reflected wave. The resulting wave is called a **standing wave.** In a standing wave, certain parts of the wave are always at the rest position because of total destructive interference between all the waves. Other parts have a large amplitude because of constructive interference.

A standing wave only *looks* as if it is standing still. Waves are actually going in both directions. Standing waves can be formed with transverse waves, such as when a musician plucks a guitar string, as well as with longitudinal waves.

✔️ **Reading Check** How can interference and reflection cause standing waves?

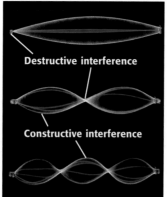

Figure 5 *When you move a rope at certain frequencies, you can create different standing waves.*

Figure 6 *A marimba produces notes through the resonance of air columns.*

ⓐ The marimba bars are struck with a mallet, causing the bars to vibrate.

ⓑ The vibrating bars cause the air in the columns to vibrate.

ⓒ The lengths of the columns have been adjusted so that the resonant frequency of the air column matches the frequency of the bar.

ⓓ The air column resonates with the bar, increasing the amplitude of the vibrations to produce a loud note.

Answer to Reading Check

A standing wave results from a wave that is reflected between two fixed points. Interference from the wave and reflected waves cause certain points to remain at rest and certain points to remain at a large amplitude.

Homework ── GENERAL

Research Resonance is a very important concept in music. Have students conduct research to learn how resonance affects the sounds produced by different musical instruments. Students can share their results by writing a report, by making a poster, or by playing a musical instrument. **LS** Logical

Resonance

The frequencies at which standing waves are made are called *resonant frequencies*. When an object vibrating at or near the resonant frequency of a second object causes the second object to vibrate, **resonance** occurs. A resonating object absorbs energy from the vibrating object and vibrates, too. An example of resonance is shown in **Figure 6** on the previous page.

You may be familiar with another example of resonance at home—in your shower. When you sing in the shower, certain frequencies create standing waves in the air that fills the shower stall. The air resonates in much the same way that the air column in a marimba does. The amplitude of the sound waves becomes greater. So your voice sounds much louder.

standing wave a pattern of vibration that simulates a wave that is standing still

resonance a phenomenon that occurs when two objects naturally vibrate at the same frequency; the sound produced by one object causes the other object to vibrate

SECTION Review

Summary

- Waves reflect after hitting a barrier.
- Refraction is the bending of a wave when it passes through different media.
- Waves bend around barriers or through openings during diffraction.
- The result of two or more waves overlapping is called interference.
- Amplitude increases during constructive interference and decreases during destructive interference.
- Resonance occurs when a vibrating object causes another object to vibrate at one of its resonant frequencies.

Using Key Terms

Complete each of the following sentences by choosing the correct term from the word bank.

   refraction   reflection
   diffraction   interference

1. ___ happens when a wave passes from one medium to another at an angle.

2. The bending of a wave around a barrier is called ___.

3. We can see the moon because of the ___ of sunlight off it.

Understanding Key Ideas

4. The combining of waves as they overlap is known as
   a. interference.
   b. diffraction.
   c. refraction.
   d. resonance.

5. Name two wave interactions that can occur when a wave encounters a barrier.

6. Explain why you can hear two people talking even after they walk around a corner.

7. Explain what happens when two waves encounter one another in destructive interference.

Critical Thinking

8. **Making Inferences** Sometimes, when music is played loudly, you can feel your body shake. Explain what is happening in terms of resonance.

9. **Applying Concepts** How could two waves on a rope interfere so that the rope did not move at all?

Interpreting Graphics

10. In the image below, what sort of wave interaction is happening?

Developed and maintained by the National Science Teachers Association

For a variety of links related to this chapter, go to www.scilinks.org

Topic: Interactions of Waves
SciLinks code: HSM0304

Cultural Awareness   GENERAL

Marimbas The marimba is an instrument of African origin that is similar to a xylophone.

CHAPTER RESOURCES

Chapter Resource File

- Section Quiz **GENERAL**
- Section Review **GENERAL**
- Vocabulary and Section Summary **GENERAL**
- Reinforcement Worksheet **BASIC**
- Critical Thinking **ADVANCED**

Wave Energy and Speed

Teacher's Notes

Time Required

One 45-minute class period

Lab Ratings

EASY ———————————→ HARD

Teacher Prep 🧪
Student Set-Up 🧪🧪
Concept Level 🧪🧪
Clean Up 🧪🧪

MATERIALS

The materials listed are for each group of 2–4 students.

Safety Caution

Remind students to review all safety cautions and icons before beginning this lab activity.

Procedure Notes

Finding enough stopwatches for all students may be difficult. Students may have watches that can serve as timers or stopwatches.

Form a Hypothesis

2. Sample answer: Waves made by a large disturbance carry more energy than waves made by a small disturbance and travel faster than waves created by a small disturbance.

Using Scientific Methods

Skills Practice Lab

OBJECTIVES

Form hypotheses about the energy and speed of waves.

Test your hypotheses by performing an experiment.

MATERIALS

- beaker, small
- newspaper
- pan, shallow, approximately 20 cm × 30 cm
- pencils (2)
- stopwatch
- water

SAFETY

Wave Energy and Speed

If you threw a rock into a pond, waves would carry energy away from the point of origin. But if you threw a large rock into a pond, would the waves carry more energy away from the point of origin than waves caused by a small rock? And would a large rock make waves that move faster than waves made by a small rock? In this lab, you'll answer these questions.

Ask a Question

1. In this lab, you will answer the following questions: Do waves made by a large disturbance carry more energy than waves made by a small disturbance? Do waves created by a large disturbance travel faster than waves created by a small disturbance?

Form a Hypothesis

2. Write a few sentences that answer the questions above.

Test the Hypothesis

3. Place the pan on a few sheets of newspaper. Using the small beaker, fill the pan with water.

4. Make sure that the water is still. Tap the surface of the water with the eraser end of one pencil. This tap represents the small disturbance. Record your observations about the size of the waves that are made and the path they take.

CHAPTER RESOURCES

Chapter Resource File

📁 • **Datasheet for Chapter Lab**
  • **Lab Notes and Answers**

Technology

💿 **Classroom Videos**
  • Lab Video

LabBook

• Wave Speed, Frequency, and Wavelength

5 Repeat step 4. This time, use the stopwatch to record the amount of time it takes for one of the waves to reach the side of the pan. Record your data.

6 Using two pencils at once, repeat steps 4 and 5. These taps represent the large disturbance. (Try to use the same amount of force to tap the water that you used with just one pencil.) Observe and record your results.

Analyze the Results

1 **Describing Events** Compare the appearance of the waves created by one pencil with that of the waves created by two pencils. Were there any differences in amplitude (wave height)?

2 **Describing Events** Compare the amount of time required for the waves to reach the side of the pan. Did the waves travel faster when two pencils were used?

Draw Conclusions

3 **Drawing Conclusions** Do waves made by a large disturbance carry more energy than waves made by a small one? Explain your answer, using your results to support your answer. (Hint: Remember the relationship between amplitude and energy.)

4 **Drawing Conclusions** Do waves made by a large disturbance travel faster than waves made by a small one? Explain your answer.

Applying Your Data

A tsunami is a giant ocean wave that can reach a height of 30 m. Tsunamis that reach land can cause injury and enormous property damage. Using what you learned in this lab about wave energy and speed, explain why tsunamis are so dangerous. How do you think scientists can predict when tsunamis will reach land?

CHAPTER RESOURCES

Workbooks

Whiz-Bang Demonstrations
• Pitch Forks **GENERAL**

Long-Term Projects & Research Ideas
• A Whale of a Wave **ADVANCED**

Assignment Guide

Section	Questions
1	1, 5, 9, 14
2	2, 4, 7, 10–12, 17
3	3, 6, 8, 13, 15–16

ANSWERS

Using Key Terms

1. A longitudinal wave is a wave that moves parallel to the direction of the disturbance of the particles. A transverse wave is a wave that moves perpendicular to the direction of the disturbance of the particles.

2. Wavelength is the distance between corresponding points on a wave. Amplitude is the maximum distance that the medium is displaced by the wave.

3. Reflection occurs when a wave strikes a barrier and reverses direction. Refraction occurs when a wave enters a different medium and its speed and direction change.

Understanding Key Ideas

4. a
5. b
6. c
7. b
8. a
9. b

Chapter Review

USING KEY TERMS

For each pair of terms, explain how the meanings of the terms differ.

1 *longitudinal* wave and *transverse wave*

2 *wavelength* and *amplitude*

3 *reflection* and *refraction*

UNDERSTANDING KEY IDEAS

Multiple Choice

4 As the wavelength increases, the frequency
  a. decreases.
  b. increases.
  c. remains the same.
  d. increases and then decreases.

5 Waves transfer
  a. matter.   c. particles.
  b. energy.   d. water.

6 Refraction occurs when a wave enters a new medium at an angle because
  a. the frequency changes.
  b. the amplitude changes.
  c. the wave speed changes.
  d. None of the above

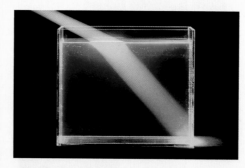

7 The wave property that is related to the height of a wave is the
  a. wavelength.
  b. amplitude.
  c. frequency.
  d. wave speed.

8 During constructive interference,
  a. the amplitude increases.
  b. the frequency decreases.
  c. the wave speed increases.
  d. All of the above

9 Waves that don't require a medium are
  a. longitudinal waves.
  b. electromagnetic waves.
  c. surface waves.
  d. mechanical waves.

Short Answer

10 Draw a transverse wave and a longitudinal wave. Label a crest, a trough, a compression, a rarefaction, and wavelengths. Also, label the amplitude on the transverse wave.

11 What is the relationship between frequency, wave speed, and wavelength?

Math Skills

12 A fisherman in a row boat notices that one wave crest passes his fishing line every 5 s. He estimates the distance between the crests to be 1.5 m and estimates that the crests of the waves are 0.5 m above the troughs. Using this data, determine the amplitude and speed of the waves.

10. Answers should include a transverse and longitudinal wave with a correctly labeled crest, trough, and amplitude in the transverse wave and compression and rarefaction in the longitudinal wave, as shown earlier in this chapter.

11. Wave speed is the frequency multiplied by the wavelength. The longer the wavelength, the lower the frequency. The shorter the wavelength, the higher the frequency.

12. $f = 1/5$ s $= 0.2$ Hz; $\lambda = 1.5$ m; $v = \lambda \times f = 1.5$ m $\times 0.2$ Hz $= 0.3$ m/s.
Amplitude $= 0.5$ m $\div 2 = 0.25$ m

CRITICAL THINKING

13 **Concept Mapping** Use the following terms to create a concept map: *wave, refraction, transverse wave, longitudinal wave, wavelength, wave speed,* and *diffraction.*

14 **Analyzing Ideas** You have lost the paddles for the canoe you rented, and the canoe has drifted to the center of a pond. You need to get it back to the shore, but you do not want to get wet by swimming in the pond. Your friend suggests that you drop rocks behind the canoe to create waves that will push the canoe toward the shore. Will this solution work? Why or why not?

15 **Applying Concepts** Some opera singers can use their powerful voices to break crystal glasses. To do this, they sing one note very loudly and hold it for a long time. While the opera singer holds the note, the walls of the glass move back and forth until the glass shatters. Explain in terms of resonance how the glass shatters.

16 **Analyzing Processes** After setting up stereo speakers in your school's music room, you notice that in certain areas of the room, the sound from the speakers is very loud. In other areas, the sound is very soft. Using the concept of interference, explain why the sound levels in the music room vary.

17 **Predicting Consequences** A certain sound wave travels through water with a certain wavelength, frequency, and wave speed. A second sound wave with twice the frequency of the first wave then travels through the same water. What is the second wave's wavelength and wave speed compared to those of the first wave?

INTERPRETING GRAPHICS

18 Look at the waves below. Rank the waves from highest energy to lowest energy, and explain your reasoning.

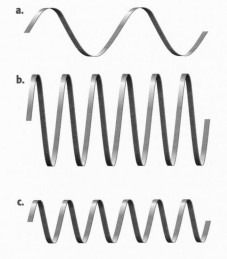

a.

b.

c.

Critical Thinking

13. An answer to this exercise can be found at the end of this book.

14. Waves carry energy, not matter. So, waves will make the canoe bob up and down in the pond but will not push the canoe closer to the shore.

15. The vibrations from the singer's voice cause the glass to vibrate at one of its resonant frequencies. It does not take much to make it vibrate with a large amplitude. But the glass is not very flexible, so it shatters.

16. With constructive interference, the crest of one wave overlaps with the crest of another wave, making a new wave with a higher crest and greater amplitude or loudness. With destructive interference, the crest of one wave overlaps with the trough of another wave, decreasing the amplitude and making the sound softer.

17. The second wave will have half the wavelength of the first wave, and the same wave speed. (Wave Speed depends on the medium, and not on wavelength or frequency.)

Interpreting Graphics

18. rank: *b, c, a*

Wave *b* has a high frequency and a large amplitude. Wave *c* has a high frequency and a small amplitude. Wave *a* has a low frequency and a small amplitude.

high frequency and large amplitude = more energy

low frequency and small amplitude = less energy

CHAPTER RESOURCES

Chapter Resource File

- Chapter Review GENERAL
- Chapter Test A GENERAL
- Chapter Test B ADVANCED
- Chapter Test C SPECIAL NEEDS
- Vocabulary Activity GENERAL

Workbooks

Study Guide
- Assessment resources are also available in Spanish.

Teacher's Note

To provide practice under more realistic testing conditions, give students 20 minutes to answer all of the questions in this Standardized Test Preparation.

MISCONCEPTION
/// **ALERT** \\\

Answers to the standardized test preparation can help you identify student misconceptions and misunderstandings.

READING

Passage 1

1. C
2. I
3. B

➕ **TEST DOCTOR**

Question 3: The passage implies that the disaster described involved the largest tsunami in the history of Kodiak, but it does not say that a tsunami had never struck Kodiak before. It also does not imply that tsunamis are common in Kodiak. The passage states the monetary cost of the damage but does not state that it caused the town to go into debt. Answer B is the most reasonable choice.

READING

Read each of the passages below. Then, answer the questions that follow each passage.

Passage 1 On March 27, 1964, a powerful earthquake rocked Alaska. The earthquake started on land near Anchorage, and the seismic waves spread quickly in all directions. The earthquake created a series of ocean waves called <u>tsunamis</u> in the Gulf of Alaska. In the deep water of the gulf, the tsunamis were short and far apart. But as these waves entered the shallow water surrounding Kodiak Island, off the coast of Alaska, they became taller and closer together. Some reached heights of nearly 30 m! The destructive forces of the earthquake and tsunamis killed 21 people and caused $10 million in damage to Kodiak, which made this marine disaster the worst in the town's 200-year history.

1. In the passage, what does *tsunami* mean?
   A a seismic wave
   B an earthquake
   C an ocean wave
   D a body of water

2. Which of these events happened first?
   F The tsunamis became closer together.
   G Tsunamis entered the shallow water.
   H Tsunamis formed in the Gulf of Alaska.
   I An earthquake began near Anchorage.

3. Which conclusion is **best** supported by information given in the passage?
   A Kodiak had never experienced a tsunami before 1964.
   B Tsunamis and an earthquake were the cause of Kodiak's worst marine disaster in 200 years.
   C Tsunamis are common in Kodiak.
   D The citizens of Kodiak went into debt after the 1964 earthquake.

Passage 2 Resonance was partially responsible for the destruction of the Tacoma Narrows Bridge, in Washington. The bridge opened in July 1940 and soon earned the nickname Galloping Gertie because of its wavelike motions. These motions were created by wind that blew across the bridge. The wind caused vibrations that were close to a resonant frequency of the bridge. Because the bridge was in resonance, it absorbed a large amount of energy from the wind, which caused it to vibrate with a large amplitude. On November 7, 1940, a supporting cable slipped, and the bridge began to twist. The twisting of the bridge, combined with high winds, further increased the amplitude of the bridge's motion. Within hours, the amplitude became so great that the bridge collapsed. Luckily, all of the people on the bridge that day were able to escape before it crashed into the river below.

1. What caused wavelike motions in the Tacoma Narrows Bridge?
   A wind that caused vibrations that were close to the resonant frequency of the bridge
   B vibrations from cars going over the bridge
   C twisting of a broken support cable
   D an earthquake

2. Why did the bridge collapse?
   F A supporting cable slipped.
   G It absorbed a great amount of energy from the wind.
   H The amplitude of the bridge's vibrations became great enough.
   I Wind blew across it.

Passage 2

1. A
2. H

 ➕ **TEST DOCTOR**

Question 2: Some students may choose answer F. Although the slippage of a supporting cable ultimately led to the bridge's collapse, the more direct cause of its collapse was when the amplitude of the bridge's vibrations became great enough.

INTERPRETING GRAPHICS

Use the figure below to answer the questions that follow.

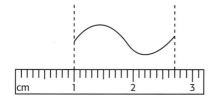

1. This wave was generated in a laboratory investigation. What is the wavelength of the wave?

A 1.5 cm

B 1.7 cm

C 2.0 cm

D 2.7 cm

2. If the frequency of the wave shown were doubled, what would the wavelength of the wave be?

F 0.85 cm

G 1.35 cm

H 3.4 cm

I 5.4 cm

3. What is the amplitude of the wave shown?

A 0.85 cm

B 1.7 cm

C 2.7 cm

D There is not enough information to determine the answer.

MATH

Read each question below, and choose the best answer.

1. How is the product of $5 \times 5 \times 5 \times 2 \times 2 \times 2 \times 2$ expressed in exponential notation?

A $3^5 \times 4^2$

B $5^3 \times 2^4$

C $5^7 \times 2^7$

D 10^7

2. Mannie purchased 8.9 kg of dog food from the veterinarian. How many grams of dog food did he purchase?

F 8,900 g

G 890 g

H 89 g

I 0.89 g

3. What is the area of a rectangle whose sides are 3 cm long and 7.5 cm long?

A 10.5 cm²

B 12 cm²

C 21 cm²

D 22.5 cm²

4. An underwater sound wave traveled 1.5 km in 1 s. How far would it travel in 4 s?

F 5.0 km

G 5.5 km

H 6.0 km

I 6.5 km

5. During a tennis game, the person serving the ball is allowed only 2 serves to start a point. Hannah plays a tennis match and is able to use 50 of her 63 first serves to start a point. What is the **best** estimate of Hannah's first-service percentage?

A 126%

B 88%

C 81.5%

D 79%

 **TEST DOCTOR**

Question 1: Students may be tempted to choose D if they simply read the marking on the ruler at 2.7 cm. The wave begins at the 1.0 cm marking, however, so its wavelength is 2.7 cm − 1.0 cm = 1.7 cm.

MATH

1. B

2. F

3. D

4. H

5. D

TEST DOCTOR

Question 5: The first sentence of this question is extraneous information. The question only asks for percentage of first serves, and how many attempts at a serve the player is allowed does not actually pertain to the question.

Standardized Test Preparation

CHAPTER RESOURCES

Chapter Resource File

• Standardized Test Preparation **GENERAL**

State Resources

For specific resources for your state, visit **go.hrw.com** and type in the keyword **HSMSTR**.

Science, Technology, and Society

Background

A radio telescope collects radio waves, just as an optical telescope collects visible light. However, to bring radio waves into sharp focus, a radio telescope must be much larger than an optical telescope because radio waves have much longer wavelengths than visible light waves do.

Scientific Discoveries

Background

Newton could not explain the curious phenomenon of colored "rings" that he noted when he placed two lenses on top of each other, because it did not fit in with his corpuscular theory of the nature of light. When revisited by Young, this interference effect of light was shown to be one of the best proofs of the wave nature of light.

Thomas Young's demonstration of the interference of light made little impression when Young announced it in 1803. It took another decade of studies and experiments by Augustin Fresnel (1788–1827) to convince the staunchest of Newton's supporters of the wave nature of light.

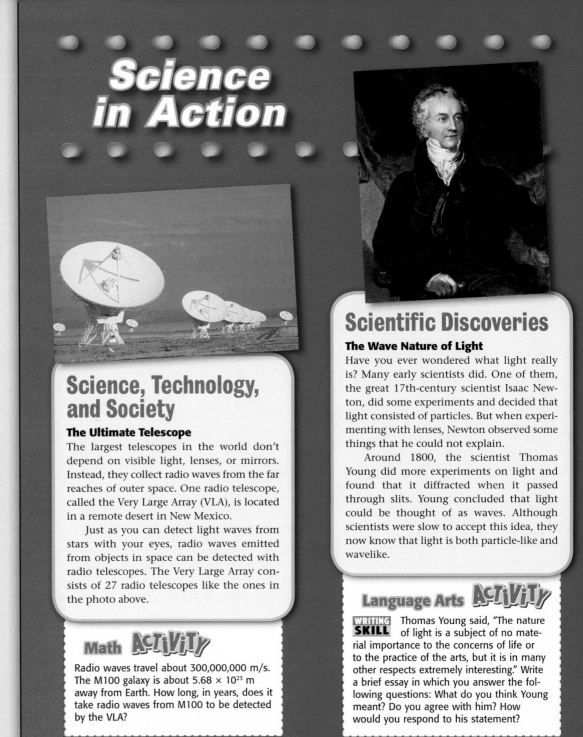

Science in Action

Science, Technology, and Society

The Ultimate Telescope

The largest telescopes in the world don't depend on visible light, lenses, or mirrors. Instead, they collect radio waves from the far reaches of outer space. One radio telescope, called the Very Large Array (VLA), is located in a remote desert in New Mexico.

Just as you can detect light waves from stars with your eyes, radio waves emitted from objects in space can be detected with radio telescopes. The Very Large Array consists of 27 radio telescopes like the ones in the photo above.

Math ACTIVITY

Radio waves travel about 300,000,000 m/s. The M100 galaxy is about 5.68×10^{23} m away from Earth. How long, in years, does it take radio waves from M100 to be detected by the VLA?

Scientific Discoveries

The Wave Nature of Light

Have you ever wondered what light really is? Many early scientists did. One of them, the great 17th-century scientist Isaac Newton, did some experiments and decided that light consisted of particles. But when experimenting with lenses, Newton observed some things that he could not explain.

Around 1800, the scientist Thomas Young did more experiments on light and found that it diffracted when it passed through slits. Young concluded that light could be thought of as waves. Although scientists were slow to accept this idea, they now know that light is both particle-like and wavelike.

Language Arts ACTIVITY

WRITING SKILL Thomas Young said, "The nature of light is a subject of no material importance to the concerns of life or to the practice of the arts, but it is in many other respects extremely interesting." Write a brief essay in which you answer the following questions: What do you think Young meant? Do you agree with him? How would you respond to his statement?

Answer to Math Activity

$d = r \times t$

$t = d \div r$

$t = 5.68 \times 10^{23}$ m $\div$ 300,000,000 m/s $=$ 1.89×10^{15} s

seconds/year = 60 s/min $\times$ 60 min/h $\times$ 24 h/day $\times$ 365 days/year = 3.15×10^7 s/year

1.89×10^{15} s $\div$ 3.15×10^7 s/year $= 6.00 \times 10^7$ (60 million) years

Answer to Language Arts Activity

Sample answer: Thomas meant that you don't need to know anything about light in order to use it and appreciate it. Nevertheless, the nature of light is very interesting. I think he was correct. Visual artists make good use of light effects, and people can see light, all without needing to worry about whether light consists of particles or waves. But the question of what light *is* is still interesting.

Estela Zavala

Ultrasonographer Estela Zavala is a registered diagnostic medical ultrasonographer who works at Austin Radiological Association in Austin, Texas. Most people have seen a picture of a sonogram showing an unborn baby inside its mother's womb. Ultrasound technologists make these images with an ultrasound machine, which sends harmless, high-frequency sound waves into the body. Zavala uses ultrasound to form images of organs in the body. Zavala says about her education, "After graduating from high school, I went to an X-ray school to be licensed as an X-ray technologist. First, I went to an intensive one-month training program. After that, I worked for a licensed radiologist for about a year. Finally, I attended a year-long ultrasound program at a local community college before becoming fully licensed." What Zavala likes best about her job is being able to help people by finding out what is wrong with them without surgery. Before ultrasound, surgery was the only way to find out about the health of someone's organs.

Social Studies ACTiViTY

WRITING SKILL Research the different ways in which ultrasound technology is used in medical practice today. Write a few paragraphs about what you learn.

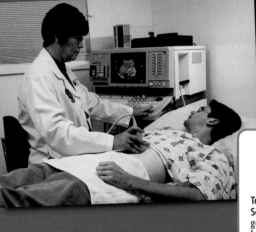

go.hrw.com
To learn more about these Science in Action topics, visit go.hrw.com and type in the keyword **HP5WAVF.**

Current Science
Check out Current Science® articles related to this chapter by visiting go.hrw.com. **Just type in the keyword HP5CS20.**

Background

When ultrasound waves pass from one material to different material, such as from muscle tissues to the lungs, some of the waves are reflected back toward the source. In addition, the speed of sound is slightly different in these different materials, so the time it takes for a reflection to arrive back at the detector depends on the material the reflection passes through. A computer is able to build an image based on the patterns of reflections detected.

Answer to Social Studies Activity

In addition to its well-known use in imaging the developing fetus in the uterus, ultrasound is also used in imaging other fluid-filled spaces in the body such as the kidney, heart, and bladder.

The Nature of Sound
Chapter Planning Guide

Compression guide:
To shorten instruction because of time limitations, omit Section 4.

OBJECTIVES	LABS, DEMONSTRATIONS, AND ACTIVITIES	TECHNOLOGY RESOURCES
PACING • 90 min pp. 28–35 **Chapter Opener**	**SE** Start-up Activity, p. 29 `GENERAL`	**OSP** Parent Letter ■ `GENERAL` **CD** Student Edition on CD-ROM **CD** Guided Reading Audio CD ■ **TR** Chapter Starter Transparency* **VID** Brain Food Video Quiz
Section 1 What Is Sound? • Describe how vibrations cause sound. • Explain how sound is transmitted through a medium. • Explain how the human ear works, and identify its parts. • Identify ways to protect your hearing.	**TE** Demonstration Vibrations of Stereo Speakers, p. 30 `GENERAL` **SE** Quick Lab Good Vibrations, p. 31 `GENERAL` **TE** Activity Sounds in Your World, p. 31 `GENERAL` **SE** Connection to Biology Vocal Sounds, p. 32 `GENERAL` **TE** Group Activity Sound Samples, p. 32 `GENERAL` **TE** Activity Diagrams of the Ear, p. 33 `BASIC` **LB** Whiz-Bang Demonstrations Hear Ye, Hear Ye* `GENERAL` **LB** Whiz-Bang Demonstrations Jingle Bells, Silent Bells* `GENERAL`	**CRF** Lesson Plans* **TR** Bellringer Transparency* **TR** Sounds from a Stereo Speaker* **TR** How the Human Ear Works* **SE** Internet Activity, p. 34 `GENERAL` **CRF** SciLinks Activity* `GENERAL` **CD** Science Tutor
PACING • 90 min pp. 36–41 **Section 2 Properties of Sound** • Compare the speed of sound in different media. • Explain how frequency and pitch are related. • Describe the Doppler effect, and give examples of it. • Explain how amplitude and loudness are related. • Describe how amplitude and frequency can be "seen" on an oscilloscope.	**TE** Activity Paper Cup Phones, p. 37 ◆ `GENERAL` **TE** Connection Activity Math, p. 37 `GENERAL` **TE** Activity Doppler Diagram, p. 38 `BASIC` **TE** Demonstration The Doppler Effect in Action, p. 38 `GENERAL` **SE** Quick Lab Sounding Board, p. 39 `GENERAL` **SE** School-to-Home Activity Decibel Levels, p. 40 `GENERAL` **SE** Skills Practice Lab Easy Listening, p. 52 `GENERAL` **LB** Whiz-Bang Demonstrations The Sounds of Time* `GENERAL`	**CRF** Lesson Plans* **TR** Bellringer Transparency* **TR** Frequency and Pitch* **TR** The Doppler Effect* **VID** Lab Videos for Physical Science **CD** Science Tutor
PACING • 45 min pp. 42–47 **Section 3 Interactions of Sound Waves** • Explain how echoes are made, and describe their use in locating objects. • List examples of constructive and destructive interference of sound waves. • Explain what resonance is.	**TE** Demonstration Sound Waves in Water, p. 42 `GENERAL` **TE** Connection Activity Language Arts, p. 43 `GENERAL` **TE** Connection Activity Real World, p. 44 `GENERAL` **TE** Activity Interference in Auditoriums, p. 45 `ADVANCED` **SE** Inquiry Lab The Speed of Sound, p. 126 `GENERAL` **SE** Skills Practice Lab Tuneful Tube, p. 127 `GENERAL` **SE** Skills Practice Lab The Energy of Sound, p. 128 `GENERAL` **LB** Whiz-Bang Demonstrations A Hot Tone* `GENERAL`	**CRF** Lesson Plans* **TR** Bellringer Transparency* **TR** *LINK TO EARTH SCIENCE* How Sonar Works* **TR** Echolocation* **CD** Science Tutor
PACING • 45 min pp. 48–51 **Section 4 Sound Quality** • Explain why different instruments have different sound qualities. • Describe how each family of musical instruments produces sound. • Explain how noise is different from music.	**TE** Demonstration Musical Instruments, p. 48 ◆ `GENERAL` **TE** Demonstration Families of Instruments, p. 49 `GENERAL` **LB** EcoLabs & Field Activities An Earful of Sounds* `BASIC` **LB** Long-Term Projects & Research Ideas The Caped Ace Flies Again* `ADVANCED` **SE** Science in Action Math, Social Studies, and Language Arts Activities, pp. 58–59 `GENERAL`	**CRF** Lesson Plans* **TR** Bellringer Transparency* **CD** Interactive Explorations CD-ROM Sound Bite! `GENERAL` **CD** Science Tutor

PACING • 90 min

CHAPTER REVIEW, ASSESSMENT, AND STANDARDIZED TEST PREPARATION

CRF Vocabulary Activity* `GENERAL`
SE Chapter Review, pp. 54–55 `GENERAL`
CRF Chapter Review* ■ `GENERAL`
CRF Chapter Tests A* ■ `GENERAL`, B* `ADVANCED`, C* `SPECIAL NEEDS`
SE Standardized Test Preparation, pp. 56–57 `GENERAL`
CRF Standardized Test Preparation* `GENERAL`
CRF Performance-Based Assessment* `GENERAL`
OSP Test Generator `GENERAL`
CRF Test Item Listing* `GENERAL`

Online and Technology Resources

Visit **go.hrw.com** for a variety of free resources related to this textbook. Enter the keyword **HP5SND.**

Holt Online Learning

Students can access interactive problem-solving help and active visual concept development with the *Holt Science and Technology* Online Edition available at **www.hrw.com.**

 Guided Reading Audio CD
Also in Spanish

A direct reading of each chapter for auditory learners, reluctant readers, and Spanish-speaking students.

 Science Tutor CD-ROM

Excellent for remediation and test practice.

SKILLS DEVELOPMENT RESOURCES	SECTION REVIEW AND ASSESSMENT	CORRELATIONS
SE Pre-Reading Activity, p. 28 `GENERAL` **OSP** Science Puzzlers, Twisters & Teasers* `GENERAL`		National Science Education Standards SAI 1, 2
CRF Directed Reading A* ■ `BASIC`, B* `SPECIAL NEEDS` **CRF** Vocabulary and Section Summary* ■ `GENERAL` **SE** Reading Strategy Prediction Guide, p. 30 `GENERAL` **TE** Inclusion Strategies, p. 31	**SE** Reading Checks, pp. 31, 32, 34 `GENERAL` **TE** Homework, p. 32 `GENERAL` **TE** Reteaching, p. 34 `BASIC` **TE** Quiz, p. 34 `GENERAL` **TE** Alternative Assessment, p. 34 `GENERAL` **SE** Section Review,* p. 35 `GENERAL` **CRF** Section Quiz* ■ `GENERAL`	SAI 1, 2; SPSP 1; PS 3a
CRF Directed Reading A* ■ `BASIC`, B* `SPECIAL NEEDS` **CRF** Vocabulary and Section Summary* ■ `GENERAL` **SE** Reading Strategy Reading Organizer, p. 36 `GENERAL` **SE** Math Practice The Speed of Sound, p. 37 `GENERAL` **CRF** Reinforcement Worksheet Doppler Dan's Dump Truck* `BASIC`	**SE** Reading Checks, pp. 37, 39, 40 `GENERAL` **TE** Reteaching, p. 40 `BASIC` **TE** Quiz, p. 40 `GENERAL` **TE** Alternative Assessment, p. 40 `GENERAL` **SE** Section Review,* p. 41 `GENERAL` **CRF** Section Quiz* `GENERAL`	UCP 1, 3; SAI 1, 2; PS 3a
CRF Directed Reading A* ■ `BASIC`, B* `SPECIAL NEEDS` **CRF** Vocabulary and Section Summary* ■ `GENERAL` **SE** Reading Strategy Paired Summarizing, p. 42 `GENERAL` **TE** Inclusion Strategies, p. 43	**SE** Reading Checks, pp. 43, 44, 46 `GENERAL` **TE** Reteaching, p. 46 `BASIC` **TE** Quiz, p. 46 `GENERAL` **TE** Alternative Assessment, p. 46 `GENERAL` **SE** Section Review,* p. 47 `GENERAL` **CRF** Section Quiz* ■ `GENERAL`	UCP 1; PS 3a
CRF Directed Reading A* ■ `BASIC`, B* `SPECIAL NEEDS` **CRF** Vocabulary and Section Summary* ■ `GENERAL` **SE** Reading Strategy Reading Organizer, p. 48 `GENERAL` **TE** Reading Strategy Prediction Guide, p. 49 `GENERAL` **CRF** Critical Thinking Worksheet The Noise Police* `ADVANCED`	**SE** Reading Checks, pp. 49, 51 `GENERAL` **TE** Homework, p. 49 `GENERAL` **TE** Reteaching, p. 50 `BASIC` **TE** Quiz, p. 50 `GENERAL` **TE** Alternative Assessment, p. 50 `GENERAL` **SE** Section Review,* p. 51 `GENERAL` **CRF** Section Quiz* ■ `GENERAL`	UCP 5

One-Stop Planner® CD-ROM

This CD-ROM package includes:
- Lab Materials QuickList Software
- Holt Calendar Planner
- Customizable Lesson Plans
- Printable Worksheets
- ExamView® Test Generator
- Interactive Teacher Edition
- Holt PuzzlePro® Resources
- Holt PowerPoint® Resources

SCI LINKS
NSTA

www.scilinks.org

Maintained by the **National Science Teachers Association.** See Chapter Enrichment pages for a complete list of topics.

Current Science®

Check out **Current Science** articles and activities by visiting the HRW Web site at go.hrw.com. Just type in the keyword **HP5CS21T**.

 Classroom Videos

- **Lab Videos** demonstrate the chapter lab.
- **Brain Food Video Quizzes** help students review the chapter material.
- **CNN Videos** bring science into your students' daily life.

Visual Resources

CHAPTER STARTER TRANSPARENCY

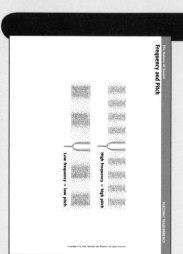

BELLRINGER TRANSPARENCIES

Section: What Is Sound?
If you've ever been near a large fireworks display, you may have *felt* the sound of the explosions. Think of other instances when you might feel sound and describe them in your **science journal.**

Section: Properties of Sound
You are the commander of a space station located about halfway between Earth and the moon. You are in the Command Center, and your chief of security tells you that sensors have just detected an explosion 61.054 km from the station. How long will it be before you hear the sound of the explosion?

Record your answer in your **science journal,** and then share your answers with the group.

TEACHING TRANSPARENCIES

Sounds from a Stereo Speaker

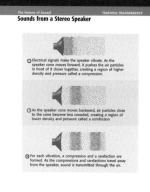

How the Human Ear Works

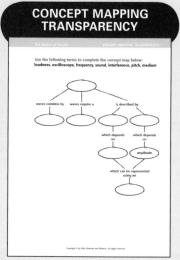

TEACHING TRANSPARENCIES

Frequency and Pitch

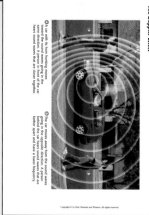

The Doppler Effect

How Sonar Works

Chapter: Exploring the Oceans

CONCEPT MAPPING TRANSPARENCY

Use the following terms to complete the concept map below:
loudness, oscilloscope, frequency, sound, interference, pitch, medium

Planning Resources

LESSON PLANS

Lesson Plan SAMPLE

Section: Waves

Pacing
Regular Schedule: with lab(s):2 days   without lab(s):2 days
Block Schedule: with lab(s):1 1/2 days   without lab(s):1 day

Objectives
1. Relate the seven properties of life to a living organism.
2. Describe seven themes that can help you to organize what you learn about biology.
3. Identify the tiny structures that make up all living organisms.
4. Differentiate between reproduction and heredity and between metabolism and homeostasis.

National Science Education Standards Covered
LSInter6:Cells have particular structures that underlie their functions.
LSMat1:Most cell functions involve chemical reactions.
LSReb1:Cells store and use information to guide their functions.
UCP1:Cell functions are regulated.
SI1: Cells can differentiate and form complete multicellular organisms.
PS1: Species evolve over time.
ESS1: The great diversity of organisms is the result of more than 3.5 billion years of evolution.
ESS2: Natural selection and its evolutionary consequences provide a scientific explanation for the fossil record of life forms as well as for the striking molecular similarities observed among the diverse species of living organisms.
ST1: The millions of different species of plants, animals, and microorganisms that live on Earth today are related by descent from common ancestors.
ST2: The energy for life primarily comes from the sun.
SPSP1: The complexity and organization of organisms accommodates the need for obtaining, transforming, transporting, releasing, and eliminating the matter and energy used to sustain the organism.
SPSP6: As matter and energy flows through different levels of organization of living systems—cells, organs, communities—and between living systems and the physical environment, chemical elements are recombined in different ways.
HNS1: Organisms have behavioral responses to internal and external stimuli.

PARENT LETTER

SAMPLE
Dear Parent,

Your son's or daughter's science class will soon begin exploring the chapter entitled "The World of Physical Science." In this chapter, students will learn about how the scientific method applies to the world of physical science and the role of physical science in the world. By the end of the chapter, students should demonstrate a clear understanding of the chapter's main ideas and be able to discuss the following topics:

1. physical science as the study of energy and matter (Section 1)
2. the role of physical science in the world around them (Section 1)
3. careers that rely on physical science (Section 1)
4. the steps used in the scientific method (Section 2)
5. examples of technology (Section 2)
6. how the scientific method is used to answer questions and solve problems (Section 2)
7. how our knowledge of science changes over time (Section 2)
8. how models represent real objects or systems (Section 3)
9. examples of different ways models are used in science (Section 3)
10. the importance of the International System of Units (Section 4)
11. the appropriate units to use for particular measurements (Section 4)
12. how area and density are derived quantities (Section 4)

Questions to Ask Along the Way
You can help your son or daughter learn about these topics by asking interesting questions such as the following:

• What are some surprising careers that use physical science?
• What is a characteristic of a good hypothesis?
• When is it a good idea to use a model?
• Why do Americans measure things in terms of inches and yards and meters ?

ALSO IN SPANISH

TEST ITEM LISTING

TEST ITEM LISTING
The World of Science   SAMPLE

MULTIPLE CHOICE
1. A limitation of models is that
   a. they are large enough to see
   b. they do not act exactly like the things that they model.
   c. they are smaller than the things that they model.
   d. they model unfamiliar things.
2. The length 10 m is equal to
   a. 100 cm.   c. 10,000 cm
   b. 1,000 cm.   d. Both (b) and (c)
   Answer: B   Difficulty: 1   Section: 3   Objective: 2
3. To be valid, a hypothesis must be
   a. testable.   c. made into a law.
   b. supported by evidence. d. Both (a) and (b)
   Answer: B   Difficulty: 1   Section: 2   Objective: 2 1
4. The statement "Sheila has a stain on her shirt" is an example of a(n)
   a. law.   c. observation.
   b. hypothesis.   d. prediction.
   Answer: B   Difficulty: 1   Section: 2   Objective: 2
5. A hypothesis is often developed out of
   a. observations.   c. laws.
   b. experiments.   d. Both (a) and (b)
   Answer: D   Difficulty: 1   Section: 2   Objective: 2
6. How many milliliters are in 3.5 kL?
   a. 3,500 mL   c. 3,500,000 mL
   b. 0.0035 mL.   d. 35,000 mL
   Answer: B   Difficulty: 1   Section: 3   Objective: 2
7. A map of Seattle is an example of a
   a. law.   c. model.
   b. theory.   d. unit.
   Answer: B   Difficulty: 1   Section: 3   Objective: 2
8. The law of conservation of mass says the
   a. mass before the total mass after the change.
   b. mass less than the total mass after the change.
   c. the same as the total mass after the change.
   d. the same as the total mass after the change.
   Answer: B   Difficulty: 1   Section: 2   Objective: 2
9. A lab has the safety icons shown below. These icons mean that you should wear
   a. only safety goggles.   c. safety goggles and a lab apron.
   b. only a lab apron.   d. safety goggles, a lab apron, and gloves.
   Answer: B   Difficulty: 1   Section: 3   Objective: 2
10. To which of the following areas might you find a geochemist at work?
    a. studying the chemistry of rocks   c. studying the ocean
    b. studying forestry   d. studying the atmosphere
    Answer: B   Difficulty: 1   Section: 3   Objective: 2

One-Stop Planner® CD-ROM

This CD-ROM includes all of the resources shown here and the following time-saving tools:

• **Lab Materials QuickList Software**
• **Customizable lesson plans**
• **Holt Calendar Planner**
• **The powerful ExamView® Test Generator**

Meeting Individual Needs

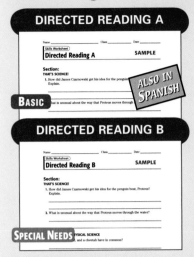

DIRECTED READING A
BASIC — ALSO IN SPANISH

DIRECTED READING B
SPECIAL NEEDS

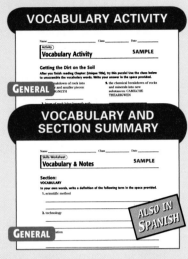

VOCABULARY ACTIVITY
GENERAL

VOCABULARY AND SECTION SUMMARY
GENERAL — ALSO IN SPANISH

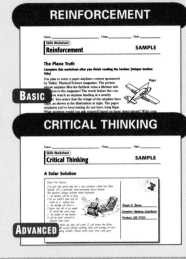

REINFORCEMENT
BASIC

CRITICAL THINKING
ADVANCED

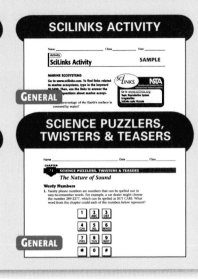

SCILINKS ACTIVITY
GENERAL

SCIENCE PUZZLERS, TWISTERS & TEASERS
GENERAL

Labs and Activities

ECOLABS & FIELD ACTIVITIES
BASIC

LONG-TERM PROJECTS & RESEARCH IDEAS
ADVANCED

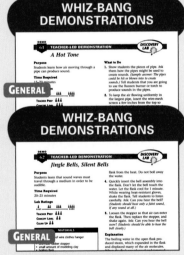

WHIZ-BANG DEMONSTRATIONS
GENERAL

WHIZ-BANG DEMONSTRATIONS
GENERAL

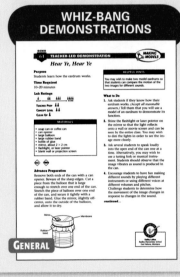

WHIZ-BANG DEMONSTRATIONS
GENERAL

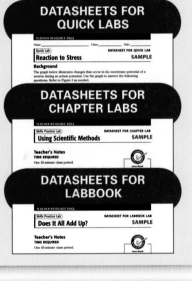

DATASHEETS FOR QUICK LABS

DATASHEETS FOR CHAPTER LABS

DATASHEETS FOR LABBOOK

Review and Assessments

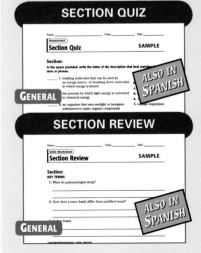

SECTION QUIZ
GENERAL — ALSO IN SPANISH

SECTION REVIEW
GENERAL — ALSO IN SPANISH

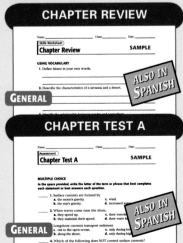

CHAPTER REVIEW
GENERAL — ALSO IN SPANISH

CHAPTER TEST A
GENERAL — ALSO IN SPANISH

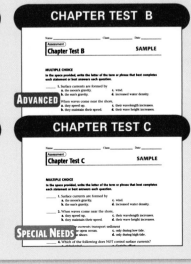

CHAPTER TEST B
ADVANCED

CHAPTER TEST C
SPECIAL NEEDS

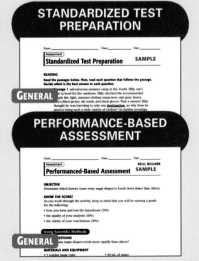

STANDARDIZED TEST PREPARATION
GENERAL

PERFORMANCE-BASED ASSESSMENT
GENERAL

This Chapter Enrichment provides relevant and interesting information to expand and enhance your presentation of the chapter material.

Section 1

What Is Sound?

Robert Boyle (1627–1691)

● Robert Boyle, a British scientist who lived in Ireland, performed a famous experiment in 1660. When a bell was suspended in a vacuum, the clapper could be seen striking the bell, but no sound was heard. This demonstration proved that sound requires a medium to propagate.

Is That a Fact!

◆ Singing sand dunes can make two very different sounds: some dunes "whistle" or "squeak" in the 500 Hz to 2,500 Hz range, but others "boom" at frequencies of 50 Hz to 300 Hz. Interestingly, the booming sands can be felt as well; the dune trembles noticeably as it booms.

◆ Only sand grains with certain sizes and shapes whistle or squeal. Sands in many locations, including beach sands in a variety of places, can be induced to whistle. Scientists are less certain what factors create booming dunes, although grains of a certain size and shape seem to be required.

Section 2

Properties of Sound

Loudness

● Loudness is expressed in decibels. An increase of 10 dB multiplies the intensity of a sound by 10 times. An increase of 20 dB is 10×10, or 100 times more intense. The human ear, however, responds logarithmically, not linearly, to intensity.

Doppler Effect

● In 1842, the Doppler effect was explained by the Austrian physicist Christian Doppler (1803–1853). He noted that there is a change in wavelength of both light and sound when either the source or the receiver (or both, if they move at different velocities) moves.

● In acoustical Doppler effects, the frequency depends on the velocity of the observer and the velocity of the source.

Is That a Fact!

◆ The longest recorded distance traveled by any audible sound in air is about 4,600 km. The volcanic explosion on the Indonesian island of Krakatau, in 1883, propelled a column of smoke and ash more than 80 km into the air. The explosion sounded like distant cannon fire to people in Australia, Singapore, and Rodriguez Island—4,600 km away in the Indian Ocean. Waves reached the Pacific coastline of Colombia 19 h later, and tsunamis were recorded in other parts of South America.

◆ Sound travels 1.7 km in about 5 s in air that is 20°C. Sound travels the same distance (1.7 km) in just over 1 s underwater and in only 0.3 s in steel.

Section 3

Interactions of Sound Waves

Bat Sonar and Human Technology

- Experiments conducted at Brown University found that bat sonar can detect the difference between echoes just 2- to 3-millionths of a second apart. The best naval sonar could differentiate between echoes about 5- to 10-millionths of a second apart.

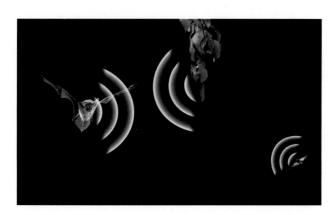

Is That a Fact!

◆ Nikola Tesla, a Croatian inventor, once created a human-made earthquake by making a steam-driven oscillator vibrate at the resonant frequency of the ground. Tesla had accurately determined the resonant frequency of the Earth! In a similar experiment, Tesla proved theories of seismic wave activity by sending waves of energy through the Earth. As these waves of energy returned, Tesla added electric current to them and thereby created a human-made bolt of lightning that measured 40 m. The accompanying thunder was heard for more than 35 km!

Section 4

Sound Quality

Recording Sound

- The two basic methods of recording sound are analog and digital recording. In analog recording, the recording medium varies continuously with the incoming signal. In digital recording, the signal is recorded as a rapid sequence of coded measurements.

Is That a Fact!

◆ Notes on a musical scale are set by exact frequencies. Although humans can hear frequencies as low as 20 Hz, the lowest frequency heard as a note is about 30 Hz. The highest frequency audible to humans is about 20,000 Hz. Middle C on the piano has a frequency of 263 Hz.

◆ Little was known about the science of sound until the 1600s. The Greeks were more interested in music than in the scientific aspects of sound. However, the Greek philosopher and mathematician Pythagoras (c. 580–500 BCE) discovered that doubling the frequency of a pitch produces a pitch one octave higher.

- Both methods preserve the varying voltage of the sound signal, but digital recording eliminates the hiss or electrical noise. Analog recordings can be improved by a noise reduction system, such as the Dolby®-system.

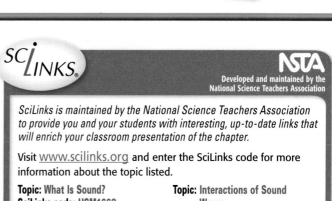

SCiLINKS.

Developed and maintained by the National Science Teachers Association

SciLinks is maintained by the National Science Teachers Association to provide you and your students with interesting, up-to-date links that will enrich your classroom presentation of the chapter.

Visit www.scilinks.org and enter the SciLinks code for more information about the topic listed.

Topic: What Is Sound?
SciLinks code: HSM1663

Topic: The Ear
SciLinks code: HSM0440

Topic: Properties of Sound
SciLinks code: HSM1233

Topic: Interactions of Sound Waves
SciLinks code: HSM0804

Topic: Sound Quality
SciLinks code: HSM1427

Overview

Tell students that this chapter will help them learn about the nature of sound. The chapter begins with a description of sound as longitudinal waves made by vibrations and carried through a medium, such as air. Properties like speed, pitch, and amplitude are then discussed. Students will then learn about interactions of sound waves, such as echoes, interference, and the Doppler effect. Differences in production of sound between musical instruments and the meaning of the term *sound quality* will then be discussed.

Assessing Prior Knowledge

Students should be familiar with the following topics:

• particles
• energy
• waves

Identifying Misconceptions

Many students wrongly intuit that sound cannot travel through solids and liquids and can travel through a vacuum and space. They also may believe that sound can be produced without any materials. Many students will also guess that hitting an object harder will change the pitch of the sound produced.

2

The Nature of Sound

About the PHOTO

Look at these dolphins swimming swiftly and silently through their watery world. Wait a minute—swiftly? Yes. Silently? No way! Dolphins use sound—clicks, squeaks, and other noises—to communicate. Dolphins also use sound to locate their food by echolocation and to find their way through murky water.

PRE-READING ACTIVITY

Graphic Organizer

Concept Map Before you read the chapter, create the graphic organizer entitled "Concept Map" described in the **Study Skills** section of the Appendix. As you read the chapter, fill in the concept map with details about each type of sound interaction.

Standards Correlations

National Science Education Standards

The following codes indicate the National Science Education Standards that correlate to this chapter. The full text of the standards can be found at the front of the book.

Chapter Opener
SAI 1, 2

Section 1 What Is Sound?
SAI 1, 2; SPSP 1; PS 3a

Section 2 Properties of Sound
UCP 1, 3; SAI 1, 2; PS 3a

Section 3 Interactions of Sound Waves
UCP 1; PS 3a

Section 4 Sound Quality
UCP 5

Chapter Lab
SAI 1, 2

Chapter Review
PS 3a

Science in Action
UCP 5; ST 2; SPSP 5

START-UP ACTIVITY

MATERIALS

FOR EACH GROUP
- pencil
- rubber band, thick
- rubber band, thin
- shoe box

Teacher's Notes:

1. Students should hear a sound and see the rubber band vibrate.

2. Students should hear a higher pitch or note from the thinner rubber band. They might not use the words *pitch* or *note* at this point, but they should be able to notice and describe the differences between the two sounds.

3. When the pencil is used, the pitch of the sound is higher.

4. A high pitch is produced by the short part of the rubber band, and a low pitch is produced by the longer part of the rubber band.

Answers

1. The thicker the rubber band is, the lower the pitch.

2. The shorter the rubber band is, the higher the pitch.

START-UP ACTIVITY

A Homemade Guitar

In this chapter, you will learn about sound. You can start by making your own guitar. It won't sound as good as a real guitar, but it will help you explore the nature of sound.

Procedure

1. Stretch a **rubber band** lengthwise around an empty **shoe box.** Place the box hollow side up. Pluck the rubber band gently. Describe what you hear.

2. Stretch **another rubber band of a different thickness** around the box. Pluck both rubber bands. Describe the differences in the sounds.

3. Put a **pencil** across the center of the box and under the rubber bands, and pluck again. Compare this sound with the sound you heard before the pencil was used.

4. Move the pencil closer to one end of the shoe box. Pluck on both sides of the pencil. Describe the differences in the sounds you hear.

Analysis

1. How did the thicknesses of the rubber bands affect the sound?

2. In steps 3 and 4, you changed the length of the vibrating part of the rubber bands. What is the relationship between the vibrating length of the rubber band and the sound that you hear?

Chapter Starter Transparency
Use this transparency to help students begin thinking about the nature of sound.

Chapter 2 • The Nature of Sound   **29**

Focus

Overview

This section introduces sound and how it is produced. Students learn how sound waves travel and how the human ear works.

 Bellringer

Tell students that if they've ever been near a large fireworks display, they may have *felt* the sound of the explosions. Ask them to think of other times they might feel sound and to describe them in their **science journal.**

Motivate

Demonstration — GENERAL

Vibrations of Stereo Speakers

Set up a stereo system with speakers—these must have woofers that reproduce low-frequency sounds—in the classroom, and remove the outer cover of the speakers. Play music that has a strong bass beat or bass notes. Students will be able to see the woofers vibrating with the bass sounds. Challenge them to explain why the vibrations of the woofers can be seen but the vibrations of the tweeters are almost imperceptible.

 Auditory/Visual

READING WARM-UP

Objectives

- Describe how vibrations cause sound.
- Explain how sound is transmitted through a medium.
- Explain how the human ear works, and identify its parts.
- Identify ways to protect your hearing.

Terms to Learn

sound wave
medium

READING STRATEGY

Prediction Guide Before reading this section, predict whether each of the following statements is true or false:

- Sound waves are made by vibrations.
- Sound waves push air particles along until they reach your ear.

What Is Sound?

You are in a restaurant, and without warning, you hear a loud crash. A waiter dropped a tray of dishes. What a mess! But why did dropping the dishes make such a loud sound?

In this section, you'll find out what causes sound and what characteristics all sounds have in common. You'll also learn how your ears detect sound and how you can protect your hearing.

Sound and Vibrations

As different as they are, all sounds have some things in common. One characteristic of sound is that it is created by vibrations. A *vibration* is the complete back-and-forth motion of an object. **Figure 1** shows one way sound is made by vibrations.

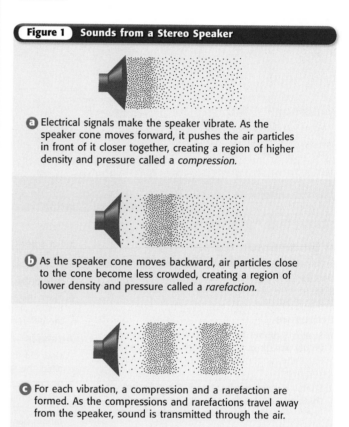

Figure 1 Sounds from a Stereo Speaker

(a) Electrical signals make the speaker vibrate. As the speaker cone moves forward, it pushes the air particles in front of it closer together, creating a region of higher density and pressure called a *compression.*

(b) As the speaker cone moves backward, air particles close to the cone become less crowded, creating a region of lower density and pressure called a *rarefaction.*

(c) For each vibration, a compression and a rarefaction are formed. As the compressions and rarefactions travel away from the speaker, sound is transmitted through the air.

CHAPTER RESOURCES

Chapter Resource File

- Lesson Plan
- Directed Reading A BASIC
- Directed Reading B SPECIAL NEEDS

Technology

 Transparencies
- Sounds from a Stereo Speaker

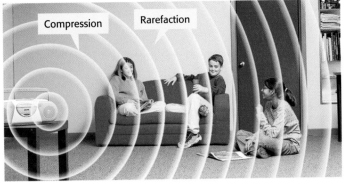

Compression   Rarefaction

Figure 2 *You can't actually see sound waves, but they can be represented by spheres that spread out in all directions.*

Sound Waves

Longitudinal (LAHN juh TOOD'n uhl) waves are made of compressions and rarefactions. A **sound wave** is a longitudinal wave caused by vibrations and carried through a substance. The particles of the substance, such as air particles, vibrate back and forth along the path that the sound wave travels. Sound is transmitted through the vibrations and collisions of the particles. Because the particles vibrate back and forth along the paths that sound travels, sound travels as longitudinal waves.

Sound waves travel in all directions away from their source, as shown in **Figure 2.** However, air or other matter does not travel with the sound waves. The particles of air only vibrate back and forth. If air did travel with sound, wind gusts from music speakers would blow you over at a school dance!

sound wave a longitudinal wave that is caused by vibrations and that travels through a material medium

✔ **Reading Check** What do sound waves consist of? (*See the Appendix for answers to Reading Checks.*)

Good Vibrations

1. Gently strike a **tuning fork** on a **rubber eraser.** Watch the prongs, and listen for a sound. Describe what you see and what you hear.
2. Lightly touch the fork with your fingers. What do you feel?
3. Grasp the prongs of the fork firmly with your hand. What happens to the sound?
4. Strike the tuning fork on the eraser again, and dip the prongs in a **cup of water.** Describe what happens to the water.
5. Record your observations.

MATERIALS

FOR EACH GROUP
- eraser, rubber
- plastic cup of water, small
- tuning fork

Safety Caution: Remind students that the tuning forks should not touch their eyes or eyeglasses.

Answers

1. Students should hear a faint sound and they may or may not see the tuning fork vibrate, depending on how hard the fork was struck.
2. Students should feel the prongs vibrate.
3. The sound will immediately stop.
4. The vibrations of the prongs will create waves in the water in the cup.

Sound in Space? Some science fiction movies are full of scenes of roaring spacecraft, loud drilling on asteroids, or deafening explosions during fictional space battles. Discuss with students why these movies are not scientifically accurate. **LS** Logical

Sound Samples Have groups of students search the school and grounds to record a variety of sounds. If possible, use a tape recorder or video camera. Instruct them to find the following sounds:

- high-pitched sound
- sound from above or overhead
- repeating sound
- sound that would startle
- irritating sound
- sound made by an animal
- sound made by wind
- sound made by something moving

Students should record what the sound is and where it was made. Compare and discuss what different groups found. **LS** Interpersonal Co-op Learning

Answer to Reading Check
Sound needs a medium in order to travel.

Figure 3 Tubing is connected to a pump that is removing air from the jar. As the air is removed, the ringing alarm clock sounds quieter and quieter.

medium a physical environment in which phenomena occur

CONNECTION TO Biology

Vocal Sounds The vibrations that produce your voice are made inside your throat. When you speak, laugh, or sing, your lungs force air up your windpipe, causing your vocal cords to vibrate.

Do some research, and find out what role different parts of your throat and mouth play in making vocal sounds. Make a poster in which you show the different parts, and explain the role they play in shaping sound waves. **ACTIVITY**

Sound and Media

Another characteristic of sound is that all sound waves require a medium (plural, *media*). A **medium** is a substance through which a wave can travel. Most of the sounds that you hear travel through air at least part of the time. But sound waves can also travel through other materials, such as water, glass, and metal.

In a vacuum, however, there are no particles to vibrate. So, no sound can be made in a vacuum. This fact helps to explain the effect described in **Figure 3**. Sound must travel through air or some other medium to reach your ears and be detected.

✓ **Reading Check** What does sound need in order to travel?

How You Detect Sound

Imagine that you are watching a suspenseful movie. Just before a door is opened, the background music becomes louder. You know that there is something scary behind that door! Now, imagine watching the same scene without the sound. You would have more difficulty figuring out what's going on if there were no sound.

Figure 4 shows how your ears change sound waves into electrical signals that allow you to hear. First, the outer ear collects sound waves. The vibrations then go to your middle ear. Very small organs increase the size of the vibrations here. These vibrations are then picked up by organs in your inner ear. Your inner ear changes vibrations into electrical signals that your brain interprets as sound.

Homework ———— GENERAL

Hearing Aids Because so many young people listen to very loud music through headphones, one of the next major industries may be the manufacturing of hearing aids. Have students research the different kinds of hearing aids available, how they work, and their costs. **LS** Logical

WEIRD SCIENCE

A cricket hears through its front legs and produces a series of chirps, or trills, by rubbing its two front wings together. And the pistol shrimp makes a sound much like a gunshot by snapping shut its enlarged claw.

Figure 4 | How the Human Ear Works

a The **outer ear** acts as a funnel for sound waves. The *pinna* collects sound waves and directs them into the *ear canal*.

b In the **middle ear,** three bones—the *hammer, anvil,* and *stirrup*—act as levers to increase the size of the vibrations.

c In the **inner ear,** vibrations created by sound are changed into electrical signals for the brain to interpret.

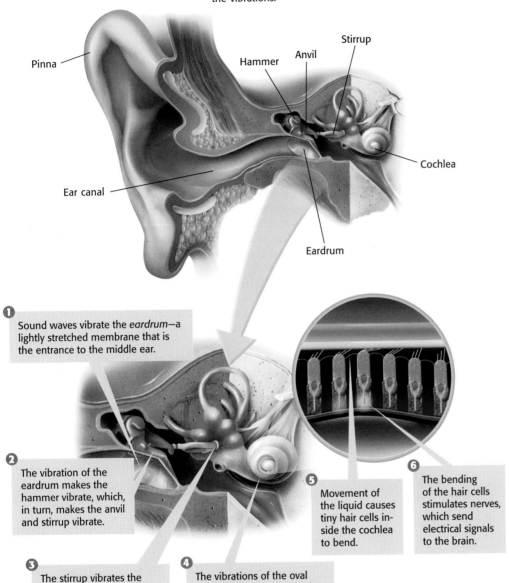

Pinna

Hammer   Anvil   Stirrup

Cochlea

Ear canal

Eardrum

1 Sound waves vibrate the *eardrum*—a lightly stretched membrane that is the entrance to the middle ear.

2 The vibration of the eardrum makes the hammer vibrate, which, in turn, makes the anvil and stirrup vibrate.

3 The stirrup vibrates the *oval window*—the entrance to the inner ear.

4 The vibrations of the oval window create waves in the liquid inside the *cochlea*.

5 Movement of the liquid causes tiny hair cells inside the cochlea to bend.

6 The bending of the hair cells stimulates nerves, which send electrical signals to the brain.

BRAIN FOOD

The inner ear contains the organs that are responsible for maintaining balance. These organs are located in a hollow region called the *vestibule.* Therefore, the sense of balance is sometimes referred to as the *vestibular sense.*

Retracing the Process of Sound and Hearing Ask students what the first step is in sound production. (a vibration) Have a student draw a very simple sketch of this on the board. Then, ask students what happens to the sound. (Sound waves [longitudinal waves] are carried through a medium.) Have another student draw a very simple sketch of this process on the board. Then, ask students what happens to the sound once it reaches your ear. (It is funneled by the outer ear, magnified by the middle ear, and changed into electrical signals by the inner ear.) Have another student draw a sketch of this process on the board. LS Visual

Quiz — GENERAL

1. Name the three bones in the ear that act as levers. (hammer, anvil, stirrup)

2. What is necessary for sound to travel from its source to a listener? (a medium through which the sound waves can travel)

Alternative Assessment — GENERAL

Hearing Problem Have students write letters to a fictional boss explaining that they are having a hearing problem at work. They should explain the problem and what corrections could be made in the workplace for the benefit of everyone who works there. LS Verbal

Figure 5 *Sound is made whether or not anyone is around to hear it.*

INTERNET ACTiViTY

For another activity related to this chapter, go to **go.hrw.com** and type in the keyword **HP5SNDW.**

Making Sound Versus Hearing Sound

Have you heard this riddle? If a tree falls in the forest and no one is around to hear it, does the tree make a sound? Think about the situation pictured in **Figure 5.** When a tree falls and hits the ground, the tree and the ground vibrate. These vibrations make compressions and rarefactions in the surrounding air. So, there would be a sound!

Making sound is separate from detecting sound. The fact that no one heard the tree fall doesn't mean that there wasn't a sound. A sound was made—it just wasn't heard.

Hearing Loss and Deafness

The many parts of the ear must work together for you to hear sounds. If any part of the ear is damaged or does not work properly, hearing loss or deafness may result.

One of the most common types of hearing loss is called *tinnitus* (ti NIET us), which results from long-term exposure to loud sounds. Loud sounds can cause damage to the hair cells and nerve endings in the cochlea. Once these hairs are damaged, they do not grow back. Damage to the cochlea or any other part of the inner ear usually results in permanent hearing loss.

People who have tinnitus often say they have a ringing in their ears. They also have trouble understanding other people and hearing the difference between words that sound alike. Tinnitus can affect people of any age. Fortunately, tinnitus can be prevented.

✓ *Reading Check* What causes tinnitus?

Is That a Fact!

Imagine the sound that is created when a giant sequoia falls! The most massive living organisms in the world are giant sequoia trees, which can grow to more than 85 m tall and can be more than 9 m in diameter at the base. Scientists estimate that some of the largest giant sequoias can have masses up to 2,500 t. Some of the oldest giant sequoias are more than 3,000 years old.

Answer to Reading Check

Tinnitus is caused by long-term exposure to loud sounds.

Protecting Your Hearing

Short exposures to sounds that are loud enough to be painful can cause hearing loss. Your hearing can also be damaged by loud sounds that are not quite painful, if you are exposed to them for long periods of time. There are some simple things you can do to protect your hearing. Loud sounds can be blocked out by earplugs. You can listen at a lower volume when you are using headphones, as in **Figure 6.** You can also move away from loud sounds. If you are near a speaker playing loud music, just move away from it. When you double the distance between yourself and a loud sound, the sound's intensity to your ears will be one-fourth of what it was before.

Figure 6 *Turning your radio down can help prevent hearing loss, especially when you use headphones.*

SECTION Review

Summary

- All sounds are generated by vibrations.
- Sounds travel as longitudinal waves consisting of compressions and rarefactions.
- Sound waves travel in all directions away from their source.
- Sound waves require a medium through which to travel. Sound cannot travel in a vacuum.
- Your ears convert sound into electrical impulses that are sent to your brain.
- Exposure to loud sounds can cause hearing damage.
- Using earplugs and lowering the volume of sounds can prevent hearing damage.

Using Key Terms

1. Use the following terms in the same sentence: *sound wave* and *medium*.

Understanding Key Ideas

2. Sound travels as
   a. transverse waves.
   b. longitudinal waves.
   c. shock waves.
   d. airwaves.

3. Which part of the ear increases the size of the vibrations of sound waves entering the ear?
   a. outer ear
   b. ear canal
   c. middle ear
   d. inner ear

4. Name two ways of protecting your hearing.

Critical Thinking

5. **Analyzing Processes** Explain why a person at a rock concert will not feel gusts of wind coming out of the speakers.

6. **Analyzing Ideas** If a meteorite crashed on the moon, would you be able to hear it on Earth? Why, or why not?

7. **Identifying Relationships** Recall the breaking dishes mentioned at the beginning of this section. Why was the sound that they made so loud?

Interpreting Graphics

Use the diagram of a wave below to answer the questions that follow.

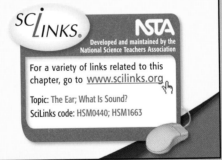

8. What kind of wave is this?

9. Draw a sketch of the diagram on a separate sheet of paper, and label the compressions and rarefactions.

10. How do vibrations make these kinds of waves?

Developed and maintained by the National Science Teachers Association

For a variety of links related to this chapter, go to www.scilinks.org

Topic: The Ear; What Is Sound?
SciLinks code: HSM0440; HSM1663

CHAPTER RESOURCES

Chapter Resource File

- Section Quiz GENERAL
- Section Review GENERAL
- Vocabulary and Section Summary GENERAL
- SciLinks Activity GENERAL
- Datasheet for Quick Lab

SECTION

2

Focus

Overview

In this section, students learn about properties of sound, what affects the speed of sound, and how pitch and frequency are related. Students also learn about the Doppler effect and how volume and amplitude are related.

Bellringer

Ask students the following question:

You are the commander of a space station located about half-way between Earth and the moon. You are in the Command Center, and your chief of security tells you that sensors have just detected an explosion 61.054 km from the station. How long will it be before you hear the sound of the explosion? (You won't hear it. Sound waves will not travel in the vacuum of space.)

READING WARM-UP

Objectives

● Compare the speed of sound in different media.

● Explain how frequency and pitch are related.

● Describe the Doppler effect, and give examples of it.

● Explain how amplitude and loudness are related.

● Describe how amplitude and frequency can be "seen" on an oscilloscope.

Terms to Learn

pitch   loudness
Doppler effect   decibel

READING STRATEGY

Reading Organizer As you read this section, create an outline of the section. Use the headings from the section in your outline.

| Table 1 Speed of Sound in Different Media ||
Medium	Speed (m/s)
Air (0°C)	331
Air (20°C)	343
Air (100°C)	366
Water (20°C)	1,482
Steel (20°C)	5,200

Properties of Sound

Imagine that you are swimming in a neighborhood pool. You can hear the high, loud laughter of small children and the soft splashing of the waves at the edge of the pool.

Why are some sounds loud, soft, high, or low? The differences between sounds depend on the properties of the sound waves. In this section, you will learn about properties of sound.

The Speed of Sound

Suppose you are standing at one end of a pool and two people from the opposite end of the pool yell at the same time. You would hear their voices at the same time. The reason is that the speed of sound depends only on the medium in which the sound is traveling. So, you would hear them at the same time—even if one person yelled louder!

How the Speed of Sound Can Change

Table 1 shows how the speed of sound varies in different media. Sound travels quickly through air, but it travels even faster in liquids and even faster in solids.

Temperature also affects the speed of sound. In general, the cooler the medium is, the slower the speed of sound. Particles of cool materials move more slowly and transmit energy more slowly than particles do in warmer materials. In 1947, pilot Chuck Yeager became the first person to travel faster than the speed of sound. Yeager flew the airplane shown in **Figure 1** at 293 m/s (about 480 mi/h) at 12,000 m above sea level. At that altitude, the temperature of the air is so low that the speed of sound is only 290 m/s.

Figure 1 *The X-1 airplane was the first vehicle to move faster than the speed of sound.*

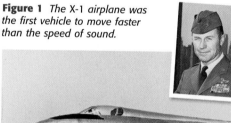

CHAPTER RESOURCES

Chapter Resource File

● **Lesson Plan**
● **Directed Reading A** BASIC
● **Directed Reading B** SPECIAL NEEDS

Technology

Transparencies
• Bellringer
• Frequency and Pitch

Is That a Fact!

In some western movies, a character would be shown putting an ear to the hard ground to find out if someone was coming. This technique actually works because sound travels faster and with less loss of energy through the ground than through air.

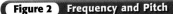

Figure 2 Frequency and Pitch

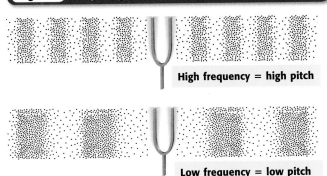

High frequency = high pitch

Low frequency = low pitch

Pitch and Frequency

How low or high a sound seems to be is the **pitch** of that sound. The *frequency* of a wave is the number of crests or troughs that are made in a given time. The pitch of a sound is related to the frequency of the sound wave, as shown in **Figure 2.** Frequency is expressed in hertz (Hz), where 1 Hz = 1 wave per second. For example, the lowest note on a piano is about 40 Hz. The screech of a bat is 10,000 Hz or higher.

✓ Reading Check What is frequency? (*See the Appendix for answers to Reading Checks.*)

Frequency and Hearing

If you see someone blow a dog whistle, the whistle seems silent to you. The reason is that the frequency of the sound wave is out of the range of human hearing. But the dog hears the whistle and comes running! **Table 2** compares the range of frequencies that humans and animals can hear. Sounds that have a frequency too high for people to hear are called *ultrasonic*.

pitch a measure of how high or low a sound is perceived to be, depending on the frequency of the sound wave

Table 2 Frequencies Heard by Different Animals	
Animal	**Frequency range (Hz)**
Bat	2,000 to 110,000
Porpoise	75 to 150,000
Cat	45 to 64,000
Beluga whale	1,000 to 123,000
Elephant	16 to 12,000
Human	20 to 20,000
Dog	67 to 45,000

The Speed of Sound

The speed of sound depends on the medium through which sound is traveling and the medium's temperature. Sound travels at 343 m/s through air at a temperature of 20°C. How far will sound travel in 20°C air in 5 s?

The speed of sound in steel at 20°C is 5,200 m/s. How far can sound travel in 5 s through steel at 20°C?

Answers to Math Practice
air: 343 m/s × 5 s = 1,715 m;
steel: 5,200 m/s × 5 s = 26,000 m

Answer to Reading Check
Frequency is the number of crests or troughs made in a given time.

Is That a Fact!
Studies now show that regions of hair cells within the inner ear are actually "tuned" to specific frequencies. Because of their differences in size and shape, hair cells have a specific resonance at which they vibrate. Thus, certain frequencies will activate some hair cells but not others.

Doppler Diagram Have students use colored pencils to draw a diagram showing the Doppler effect. Have them include a listener, a vehicle, and lines depicting sound waves. **LS Visual**

Demonstration —— **GENERAL**

The Doppler Effect in Action

Find a noisemaker that produces a constant sound (a small buzzer or something similar). Attach the noisemaker to the end of a string or rope approximately 70 cm to 90 cm long. (It may be necessary to do this demonstration outside.) Let students hear the noise from the noisemaker. Stand away from the students and away from any walls. Swing the noisemaker by the string in a wide circle over your head. Ask students to explain what they hear in terms of the Doppler effect. (When the noisemaker is coming toward them, they will hear a higher pitch because the sound waves are closer together. When it is moving away from them, they will hear a lower pitch because the sound waves are farther apart.)
LS Auditory

Figure 3 The Doppler Effect

a A car with its horn honking moves toward the sound waves going in the same direction. A person in front of the car hears sound waves that are closer together.

b The car moves away from the sound waves going in the opposite direction. A person behind the car hears sound waves that are farther apart and have a lower frequency.

Doppler effect an observed change in the frequency of a wave when the source or observer is moving

The Doppler Effect

Have you ever been passed by a car with its horn honking? If so, you probably noticed the sudden change in pitch—sort of an *EEEEEEOOooooowwn* sound—as the car went past you. The pitch you heard was higher as the car moved toward you than it was after the car passed. This higher pitch was a result of the Doppler effect. For sound waves, the **Doppler effect** is the apparent change in the frequency of a sound caused by the motion of either the listener or the source of the sound. **Figure 3** shows how the Doppler effect works.

In a moving sound source, such as a car with its horn honking, sound waves that are moving forward are going the same direction the car is moving. As a result, the compressions and rarefactions of the sound wave will be closer together than they would be if the sound source was not moving. To a person in front of the car, the frequency and pitch of the sound seem high. After the car passes, it is moving in the opposite direction that the sound waves are moving. To a person behind the car, the frequency and pitch of the sound seem low. The driver always hears the same pitch because the driver is moving with the car.

CONNECTION to
History ——————— **GENERAL**

Doppler Effect The Doppler effect is named after the Austrian mathematician Christian Doppler (1803–1853), who first proposed it in 1842 in a paper describing the colored light of double stars. In 1845, Doppler applied his theory to sound waves and tested it with trumpet players on a train. In 1929, astronomer Edwin Hubble (1889–1953) used Doppler's theory to interpret his measurements and show that the farther away from Earth a galaxy is, the faster it is moving away from Earth and the more the light from that galaxy is "shifted" toward the red end of the spectrum.

Loudness and Amplitude

If you gently tap a drum, you will hear a soft rumbling. But if you strike the drum with a large force, you will hear a much louder sound! By changing the force you use to strike the drum, you change the loudness of the sound that is created. **Loudness** is a measure of how well a sound can be heard.

Energy and Vibration

Look at **Figure 4.** The harder you strike a drum, the louder the boom. As you strike the drum harder, you transfer more energy to the drum. The drum moves with a larger vibration and transfers more energy to the air around it. This increase in energy causes air particles to vibrate farther from their rest positions.

Increasing Amplitude

When you strike a drum harder, you are increasing the amplitude of the sound waves being made. The *amplitude* of a wave is the largest distance the particles in a wave vibrate from their rest positions. The larger the amplitude, the louder the sound. And the smaller the amplitude, the softer the sound. One way to increase the loudness of a sound is to use an amplifier, shown in **Figure 5.** An amplifier receives sound signals in the form of electric current. The amplifier then increases the energy and makes the sound louder.

Reading Check What is the relationship between the amplitude of a sound and its energy of vibration?

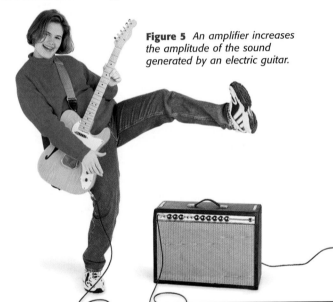

Figure 5 An amplifier increases the amplitude of the sound generated by an electric guitar.

Figure 4 When a drum is struck hard, it vibrates with a lot of energy, making a loud sound.

loudness the extent to which a sound can be heard

Sounding Board

1. With one hand, hold a **ruler** on your **desk** so that one end of it hangs over the edge.

2. With your other hand, pull the free end of the ruler up a few centimeters, and let go.

3. Try pulling the ruler up different distances. How does the distance affect the sounds you hear? What property of the sound wave are you changing?

4. Change the length of the part that hangs over the edge. What property of the sound wave is affected? Record your answers and observations.

Quick Lab

Close

Reteaching — BASIC

Sound Relationships Write the words *frequency* and *amplitude* in one line on the board, and *loudness* and *pitch* on the next line. Ask students to match each term with its counterpart. Then, ask students what kind of relationship is involved in each pair. (Frequency is related to pitch, and amplitude is related to loudness. Both are direct relationships, meaning if one increases, the other also increases.) **LS** Visual/Logical

Quiz — GENERAL

1. Name three properties of sound. (speed, pitch, and loudness)

2. Explain how a person can observe the Doppler effect. (When the source of a loud noise is moving toward or away from a listener, the sound appears to change pitch because the sound waves are moving closer together or farther apart relative to the listener's position.)

Alternative Assessment — GENERAL

Concept Mapping Have students create a concept map showing the properties of sound. Concept maps should include the vocabulary terms from this section and examples illustrating the terms. **LS** Visual

WRITING SKILL **Decibel Levels** With a parent, listen for the normal sounds that happen around your house. In your **science journal,** write down some sounds and what you think their decibel levels might be. Then, move closer to the source of each sound, and write what you think the new decibel level of each is.

decibel the most common unit used to measure loudness (symbol, dB)

Table 3 Decibel Levels of Common Sounds	
Decibel level	**Sound**
0	the softest sounds you can hear
20	whisper
25	purring cat
60	normal conversation
80	lawn mower, vacuum cleaner, truck traffic
100	chain saw, snowmobile
115	sandblaster, loud rock concert, automobile horn
120	threshold of pain
140	jet engine 30 m away
200	rocket engine 50 m away

Measuring Loudness

The most common unit used to express loudness is the **decibel** (dB). The softest sounds an average human can hear are at a level of 0 dB. Sounds that are at 120 dB or higher can be painful. **Table 3** shows some common sounds and their decibel levels.

"Seeing" Amplitude and Frequency

Sound waves are invisible. However, technology can provide a way to "see" sound waves. A device called an *oscilloscope* (uh SIL uh SKOHP) can graph representations of sound waves, as shown in **Figure 6.** Notice that the graphs look like transverse waves instead of longitudinal waves.

✓ **Reading Check** What does an oscilloscope do?

Figure 6 "Seeing" Sounds

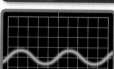  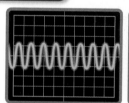 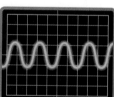

The graph on the right has a **larger amplitude** than the graph on the left. So, the sound represented on the right is **louder** than the one represented on the left.

The graph on the right has a **lower frequency** than the one on the left. So, the sound represented on the right has a **lower pitch** than the one represented on the left.

Answer to Reading Check
An oscilloscope turns sounds into electrical signals and graphs the signals.

From Sound to Electrical Signal

An oscilloscope is shown in **Figure 7.** A microphone is attached to the oscilloscope and changes a sound wave into an electrical signal. The electrical signal is graphed on the screen in the form of a wave. The graph shows the sound as if it were a transverse wave. So, the sound's amplitude and frequency are easier to see. The highest points (crests) of these waves represent compressions, and the lowest points (troughs) represent rarefactions. By looking at the displays on the oscilloscope, you can quickly see the differences in amplitude and frequency of different sound waves.

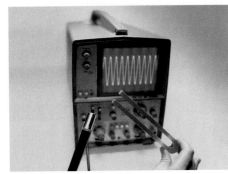

Figure 7 *An oscilloscope can be used to represent sounds.*

SECTION Review

Summary

- The speed of sound depends on the medium and the temperature.
- The pitch of a sound becomes higher as the frequency of the sound wave becomes higher. Frequency is expressed in units of Hertz (Hz), which is equivalent to waves per second.
- The Doppler effect is the apparent change in frequency of a sound caused by the motion of either the listener or the source of the sound.
- Loudness increases with the amplitude of the sound. Loudness is expressed in decibels.
- The amplitude and frequency of a sound can be measured electronically by an oscilloscope.

Using Key Terms

1. In your own words, write a definition for the term *pitch*.

2. Use the following terms in the same sentence: *loudness* and *decibel*.

Understanding Key Ideas

3. At the same temperature, in which medium does sound travel fastest?
   a. air
   b. liquid
   c. solid
   d. It travels at the same speed through all media.

4. In general, how does the temperature of a medium affect the speed of sound through that medium?

5. What property of waves affects the pitch of a sound?

6. How does an oscilloscope allow sound waves to be "seen"?

Math Skills

7. You see a distant flash of lightning, and then you hear a thunderclap 2 s later. The sound of the thunder moves at 343 m/s. How far away was the lightning?

8. In water that is near 0°C, a submarine sends out a sonar signal (a sound wave). The signal travels 1500 m/s and reaches an underwater mountain in 4 s. How far away is the mountain?

Critical Thinking

9. **Analyzing Processes** Will a listener notice the Doppler effect if both the listener and the source of the sound are traveling toward each other? Explain your answer.

10. **Predicting Consequences** A drum is struck gently, then is struck harder. What will be the difference in the amplitude of the sounds made? What will be the difference in the frequency of the sounds made?

SCILINKS.
Developed and maintained by the National Science Teachers Association

For a variety of links related to this chapter, go to www.scilinks.org

Topic: Properties of Sound
SciLinks code: HSM1233

Answers to Section Review

1. Sample answer: how low or high a sound seems, which depends on the frequency of the sound waves

2. Loudness is expressed scientifically in units of decibels.

3. c

4. Sound tends to travel fastest in higher-temperature media.

5. Pitch is determined by the frequency of sound waves.

6. An oscilloscope changes sound waves into electronic signals, which are represented on a screen.

7. 343 m/s × 2 s = 686 m

8. 1,500 m/s × 4 s = 6,000 m

9. yes; The sound waves in front of the source will become closer together as the source moves forward. Also, the listener will "meet" the sound waves more rapidly by moving toward the source. The movements of both the source and the listener will make the pitch of the sound higher.

10. The amplitude of the sound will be higher when the drum is struck harder. The frequency will not change.

```
CHAPTER RESOURCES

Chapter Resource File

• Section Quiz GENERAL
• Section Review GENERAL
• Vocabulary and Section Summary GENERAL
• Reinforcement Worksheet BASIC
• Datasheet for Quick Lab
```

Overview

In this section students learn about reflection and echolocation. They also learn about constructive and destructive wave interference and resonance.

⏰ Bellringer

Put these questions on the board:

- On an oscilloscope, does a wave with a larger amplitude (greater crests and troughs) indicate louder sound or higher pitch? (louder sound)

- As frequency increases, does pitch get higher or lower? (higher)

- What is the speed of sound dependent on? (Sample answer: the medium and its temperature)

- What do you think happens when two sound waves interact with each other? (Answers may vary.)

Motivate

Demonstration — GENERAL

Sound Waves in Water

Fill a shallow pan with water. Let it sit until the surface of the water becomes still. Tap a tuning fork, and touch the fork to the surface of the water. Ask students to explain in their **science journal** why the water moves.

LS Visual

READING WARM-UP

Objectives

- Explain how echoes are made, and describe their use in locating objects.
- List examples of constructive and destructive interference of sound waves.
- Explain what resonance is.

Terms to Learn

echo	sonic boom
echolocation	standing wave
interference	resonance

READING STRATEGY

Paired Summarizing Read this section silently. In pairs, take turns summarizing the material. Stop to discuss ideas that seem confusing.

echo a reflected sound wave

Interactions of Sound Waves

Have you ever heard of a sea canary? It's not a bird! It's a whale! Beluga whales are sometimes called sea canaries because of the many different sounds they make.

Dolphins, beluga whales, and many other animals that live in the sea use sound to communicate. Beluga whales also rely on reflected sound waves to find fish, crabs, and shrimp to eat. In this section, you will learn about reflection and other interactions of sound waves. You will also learn how bats, dolphins, and whales use sound to find food.

Reflection of Sound Waves

Reflection is the bouncing back of a wave after it strikes a barrier. You're probably already familiar with a reflected sound wave, otherwise known as an **echo.** The strength of a reflected sound wave depends on the reflecting surface. Sound waves reflect best off smooth, hard surfaces. Look at **Figure 1.** A shout in an empty gymnasium can produce an echo, but a shout in an auditorium usually does not.

The difference is that the walls of an auditorium are usually designed so that they absorb sound. If sound waves hit a flat, hard surface, they will reflect back. Reflection of sound waves doesn't matter much in a gymnasium. But you don't want to hear echoes while listening to a musical performance!

Figure 1 Sound Reflection and Absorption

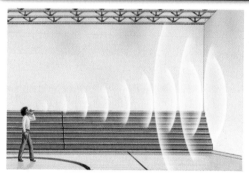 

Sound waves easily reflect off the smooth, hard walls of a gymnasium. For this reason, you hear an echo.

In well-designed auditoriums, echoes are reduced by soft materials that absorb sound waves and by irregular shapes that scatter sound waves.

CHAPTER RESOURCES

Chapter Resource File

- **Lesson Plan**
- **Directed Reading A** BASIC
- **Directed Reading B** SPECIAL NEEDS

Technology

Transparencies
- Bellringer
- *LINK TO EARTH SCIENCE* How Sonar Works
- Echolocation

Is That a Fact!

According to Greek mythology, echoes originated when the angry goddess Hera placed a curse on the wood nymph Echo that caused her to repeat whatever was said to her.

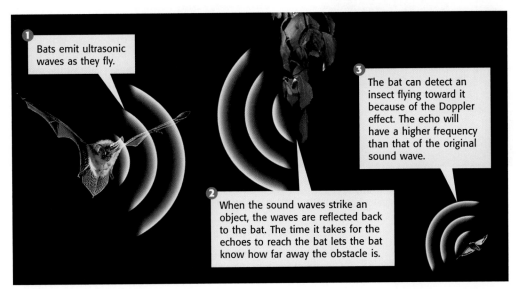

① Bats emit ultrasonic waves as they fly.

③ The bat can detect an insect flying toward it because of the Doppler effect. The echo will have a higher frequency than that of the original sound wave.

② When the sound waves strike an object, the waves are reflected back to the bat. The time it takes for the echoes to reach the bat lets the bat know how far away the obstacle is.

Figure 2 *Bats use echolocation to navigate around barriers and to find insects to eat.*

Echolocation

Beluga whales use echoes to find food. The use of reflected sound waves to find objects is called **echolocation.** Other animals—such as dolphins, bats, and some kinds of birds—also use echolocation to hunt food and to find objects in their paths. **Figure 2** shows how echolocation works. Animals that use echolocation can tell how far away something is based on how long it takes sound waves to echo back to their ears. Some animals, such as bats, also make use of the Doppler effect to tell if another moving object, such as an insect, is moving toward it or away from it.

echolocation the process of using reflected sound waves to find objects; used by animals such as bats

✔ **Reading Check** How is echolocation useful to some animals? (*See the Appendix for answers to Reading Checks.*)

Echolocation Technology

People use echoes to locate objects underwater by using sonar (which stands for **so**und **na**vigation and **r**anging). *Sonar* is a type of electronic echolocation. **Figure 3** shows how sonar works. Ultrasonic waves are used because their short wavelengths give more details about the objects they reflect off. Sonar can also help navigators on ships avoid icebergs and can help oceanographers map the ocean floor.

Figure 3 *A fish finder sends ultrasonic waves down into the water. The time it takes for the echo to return helps determine the location of the fish.*

Answer to Reading Check
Echolocation helps some animals find food.

The sound transmission class (STC) rating tells how well building materials insulate sound. The higher the number, the more the material blocks sound.

STC #	What can be heard
25	Normal speech can be heard.
30	Loud speech can be heard.
35	Loud speech is audible but not intelligible.
42	Loud speech is audible as a murmur.
45	A person must strain to hear loud speech.
48	Some loud speech is barely audible.
50	Loud speech is inaudible.

Have students research what sort of building materials have the STC ratings given above. Then, have students design a house and use this rating system to determine what building materials to use. Ask them to explain why they put the materials where they did. **LS** Logical

Answer to Reading Check

Sound wave interference can be either constructive or destructive.

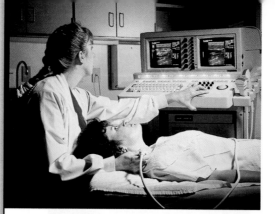

Figure 4 *Images created by ultrasonography are fuzzy, but they are a safe way to see inside a patient's body.*

interference the combination of two or more waves that results in a single wave

sonic boom the explosive sound heard when a shock wave from an object traveling faster than the speed of sound reaches a person's ears

Ultrasonography

Ultrasonography (UHL truh soh NAHG ruh fee) is a medical procedure that uses echoes to "see" inside a patient's body without doing surgery. A special device makes ultrasonic waves with a frequency that can be from 1 million to 10 million hertz, which reflect off the patient's internal organs. These echoes are then changed into images that can be seen on a television screen, as shown in **Figure 4**. Ultrasonography is used to examine kidneys, gallbladders, and other organs. It is also used to check the development of an unborn baby in a mother's body. Ultrasonic waves are less harmful to human tissue than X rays are.

Interference of Sound Waves

Sound waves also interact through interference. **Interference** happens when two or more waves overlap. **Figure 5** shows how two sound waves can combine by both constructive and destructive interference.

Orchestras and bands make use of constructive interference when several instruments of the same kind play the same notes. Interference of the sound waves causes the combined amplitude to increase, resulting in a louder sound. But destructive interference may keep some members of the audience from hearing the concert well. In certain places in an auditorium, sound waves reflecting off the walls interfere destructively with the sound waves from the stage.

✓ *Reading Check* What are the two kinds of sound wave interference?

Figure 5 **Constructive and Destructive Interference**

Sound waves from two speakers producing sound of the same frequency combine by both constructive and destructive interference.

Constructive Interference
As the compressions of one wave overlap the compressions of another wave, the sound will be louder because the amplitude is increased.

Destructive Interference
As the compressions of one wave overlap the rarefactions of another wave, the sound will be softer because the amplitude is decreased.

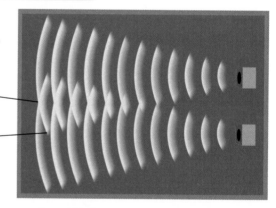

Patients who have kidney stones can have them removed painlessly with an ultrasonic procedure called *lithotripsy*. Ultrasonography is used to examine the liver, spleen, pancreas, stomach, esophagus, large intestine, small intestine, colon, heart, cranial arteries, and eyes. Dentists use ultrasonic devices to remove plaque from their patients' teeth. Veterinarians can also use ultrasound to examine and treat their animal patients.

Is That a Fact!

Mach number is not a speed. It is the ratio between the speed of an object, usually an airplane, and the speed of sound in the medium in which the object is traveling. A plane traveling at Mach 3.0 is traveling at 3 times whatever the speed of sound is at the plane's altitude.

Interference and the Sound Barrier

As the source of a sound—such as a jet plane—gets close to the speed of sound, the sound waves in front of the jet plane get closer and closer together. The result is constructive interference. **Figure 6** shows what happens as a jet plane reaches the speed of sound.

For the jet in **Figure 6** to go faster than the speed of sound, the jet must overcome the pressure of the compressed sound waves. **Figure 7** shows what happens as soon as the jet reaches supersonic speeds—speeds faster than the speed of sound. At these speeds, the sound waves trail off behind the jet. At their outer edges, the sound waves combine by constructive interference to form a *shock wave.*

A **sonic boom** is the explosive sound heard when a shock wave reaches your ears. Sonic booms can be so loud that they can hurt your ears and break windows. They can even make the ground shake as it does during an earthquake.

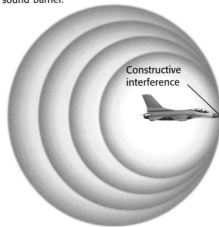

Figure 6 *When a jet plane reaches the speed of sound, the sound waves in front of the jet combine by constructive interference. The result is a high-density compression that is called the sound barrier.*

Constructive interference

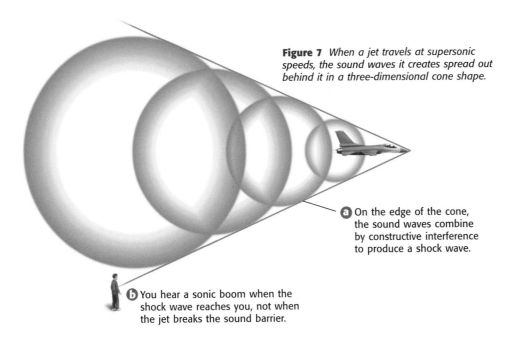

Figure 7 *When a jet travels at supersonic speeds, the sound waves it creates spread out behind it in a three-dimensional cone shape.*

a On the edge of the cone, the sound waves combine by constructive interference to produce a shock wave.

b You hear a sonic boom when the shock wave reaches you, not when the jet breaks the sound barrier.

BRAIN FOOD

The speed of sound is about 343 m/s at sea level at 20°C and is known as *Mach 1.* As a plane passes through the sound barrier created at this speed, the resulting shock wave increases the drag on the plane. The plane has to be equipped to control this change in airflow.

Is That a Fact!

A double sonic boom occurs when the space shuttle enters the atmosphere. Whenever a craft exceeds Mach 1, one shock wave is formed at the nose, and another shock wave is formed at the tail. If the shock waves are more than 0.10 s apart, you hear two sonic booms.

Close

Reteaching — BASIC

Brainstorming Section Concepts
Ask students what the two main ideas in this section are. (reflection and interference of sound waves) If they mention resonance, ask them what more general concept it goes under. (interference) Then, have them name subtopics covered in this chapter and which of the two main ideas each is related to. (reflection: echoes, echolocation, ultrasonography; interference: constructive interference, destructive interference, sound barrier, sonic boom, standing waves, resonance)
LS Logical

Quiz — GENERAL

1. Why does everything seem so quiet after a snowfall? (Snow does not reflect sound waves; it absorbs them.)

2. Have you ever sung in the shower? Why does your voice sound so much better there? (The hard, smooth walls of the shower reflect sound waves, and the interactions of the waves make your voice sound fuller.)

Alternative Assessment — GENERAL

Sound Stories Have students write fictional articles for a tabloid newspaper about a strange phenomenon caused by an interaction of sound waves they have learned about in this chapter. The science must be accurate, but they can exaggerate the effects. **LS** Verbal

---

standing wave a pattern of vibration that simulates a wave that is standing still

resonance a phenomenon that occurs when two objects naturally vibrate at the same frequency; the sound produced by one object causes the other object to vibrate

Figure 9 *When struck, a tuning fork can make another object vibrate if they both have the same resonant frequency.*

Figure 8 Resonant Frequencies of a Plucked String

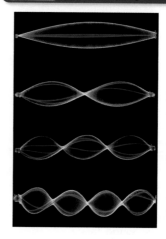

The lowest resonant frequency is called the *fundamental*.

Higher resonant frequencies are called *overtones*. The first overtone is twice the frequency of the fundamental.

The second overtone is 3 times the fundamental.

The third overtone is 4 times the fundamental.

Interference and Standing Waves

When you play a guitar, you can make some pleasing sounds, and you might even play a tune. But have you ever watched a guitar string after you've plucked it? You may have noticed that the string vibrates as a standing wave. A **standing wave** is a pattern of vibration that looks like a wave that is standing still. Waves and reflected waves of the same frequency are going through the string. Where you see maximum amplitude, waves are interfering constructively. Where the string seems to be standing still, waves are interfering destructively.

Although you can see only one standing wave, which is at the *fundamental* frequency, the guitar string actually creates several standing waves of different frequencies at the same time. The frequencies at which standing waves are made are called *resonant frequencies*. Resonant frequencies and the relationships between them are shown in **Figure 8.**

✓ Reading Check What is a standing wave?

Resonance

If you have a tuning fork, shown in **Figure 9,** that vibrates at one of the resonant frequencies of a guitar string, you can make the string make a sound without touching it. Strike the tuning fork, and hold it close to the string. The string will start to vibrate and produce a sound.

Using the vibrations of the tuning fork to make the string vibrate is an example of resonance. **Resonance** happens when an object vibrating at or near a resonant frequency of a second object causes the second object to vibrate.

MISCONCEPTION ///ALERT

Fundamentals and Harmonics When overtones are exact multiples of the fundamental, they are often called *harmonics*. However, confusion sometimes results because harmonics are numbered to include the fundamental, but overtones are not. As a result, the first overtone is the second harmonic, the second overtone is the third harmonic, and so on.

Answer to Reading Check
A standing wave is a pattern of vibration that looks like a wave that is standing still.

Resonance in Musical Instruments

Musical instruments use resonance to make sound. In wind instruments, vibrations are caused by blowing air into the mouthpiece. The vibrations make a sound, which is amplified when it forms a standing wave inside the instrument.

String instruments also resonate when they are played. An acoustic guitar, such as the one shown in **Figure 10,** has a hollow body. When the strings vibrate, sound waves enter the body of the guitar. Standing waves form inside the body of the guitar, and the sound is amplified.

Figure 10 *The body of a guitar resonates when the guitar is strummed.*

SECTION Review

Summary

- Echoes are reflected sound waves.
- Some animals can use echolocation to find food or to navigate around objects.
- People use echolocation technology in many underwater applications.
- Ultrasonography uses sound reflection for medical applications.
- Sound barriers and shock waves are created by interference.
- Standing waves form at an object's resonant frequencies.
- Resonance happens when a vibrating object causes a second object to vibrate at one of its resonant frequencies.

Using Key Terms

1. Use the following terms in the same sentence: *echo* and *echolocation.*

Complete each of the following sentences by choosing the correct term from the word bank.

interference   standing wave
sonic boom   resonance

2. When you pluck a string on a musical instrument, a(n) _____ forms.

3. When a vibrating object causes a nearby object to vibrate, _____ results.

Understanding Key Ideas

4. What causes an echo?
   a. reflection
   b. resonance
   c. constructive interference
   d. destructive interference

5. Describe a place in which you would expect to hear echoes.

6. How do bats use echoes to find insects to eat?

7. Give one example each of constructive and destructive interference of sound waves.

Math Skills

8. Sound travels through air at 343 m/s at 20°C. A bat emits an ultrasonic squeak and hears the echo 0.05 s later. How far away was the object that reflected it? (Hint: Remember that the sound must travel *to* the object and *back to* the bat.)

Critical Thinking

9. **Applying Concepts** Your friend is playing a song on a piano. Whenever your friend hits a certain key, the lamp on top of the piano rattles. Explain why the lamp rattles.

10. **Making Comparisons** Compare sonar and ultrasonography in locating objects.

For a variety of links related to this chapter, go to www.scilinks.org

Topic: Interactions of Sound Waves
SciLinks code: HSM0804

SC/LINKS® NSTA Developed and maintained by the National Science Teachers Association

Answers to Section Review

1. Sample answer: Some animals use echolocation to "see" in the dark by listening for the echoes of the sounds they make off reflected objects.

2. standing wave

3. resonance

4. a

5. Sample answer: a room with smooth walls

6. They use echolocation to find insects by emitting sounds and listening for the echoes, which tell them how far away the insect is and how fast it is moving.

7. Sample answer: constructive interference: sonic boom; destructive interference: a "dead spot" in a concert hall

8. 343 m/s × 0.05 s = 17 m
   17 m ÷ 2 = 8.5 m

9. Resonance is occurring, because the lamp on top of the piano has a resonant frequency equal to one of the notes being played.

10. Both sonar and ultrasonography make use of sound reflection to locate objects. Ultrasonography, however, involves much higher frequency sound waves than used in sonar and allows imaging of objects instead of just locating them.

CHAPTER RESOURCES

Chapter Resource File

- Section Quiz GENERAL
- Section Review GENERAL
- Vocabulary and Section Summary GENERAL

SECTION
4

Focus

Overview

This section defines *sound quality* and explains how the three main musical-instrument families (wind, percussion, and string) produce sound. Students learn the difference between music and noise.

Bellringer

Ask the following questions:

• Which strings on a piano have lower pitch? (the longer and thicker strings)

• Why does a tuba have a lower pitch than a trumpet? (Sample answer: A tuba has a longer and larger air column.)

• Why are some sounds pleasing to hear and some sounds not? Explain your answer. (Accept all reasonable answers.)

Demonstration —— GENERAL

Musical Instruments Ask two or three different volunteers who play musical instruments to demonstrate them for the class. Without discussing the different families of instruments, have students compare the sound quality of instruments within a family and from different families. Tell the band or orchestra members that they will have a chance to perform later in the lesson. LS Auditory

READING WARM-UP

Objectives

● Explain why different instruments have different sound qualities.

● Describe how each family of musical instruments produces sound.

● Explain how noise is different from music.

Terms to Learn
sound quality
noise

READING STRATEGY

Reading Organizer As you read this section, make a table comparing the way different instruments produce sound.

Sound Quality

Have you ever been told that the music you really like is just a lot of noise? If you have, you know that people can disagree about the difference between noise and music.

You might think of noise as sounds you don't like and music as sounds that are pleasant to hear. But the difference between music and noise does not depend on whether you like the sound. The difference has to do with sound quality.

What Is Sound Quality?

Imagine that the same note is played on a piano and on a violin. Could you tell the instruments apart without looking? The notes played have the same frequency. But you could probably tell them apart because the instruments make different sounds. The notes sound different because a single note on an instrument actually comes from several different pitches: the fundamental and several overtones. The result of the combination of these pitches is shown in **Figure 1.** The result of several pitches mixing together through interference is **sound quality.** Each instrument has a unique sound quality. **Figure 1** also shows how the sound quality differs when two instruments play the same note.

Figure 1 *Each instrument has a unique sound quality that results from the particular blend of overtones that it has.*

Fundamental

First overtone

Second overtone

Resulting sound

Piano

Violin

CHAPTER RESOURCES

CHAPTER RESOURCES

Chapter Resource File

• Lesson Plan
• Directed Reading A BASIC
• Directed Reading B SPECIAL NEEDS

Technology

 Transparencies
• Bellringer

Is That a Fact!

Many cultures write and play music on a basic eight-tone scale, although the notes or tones may vary greatly from scale to scale. Other cultures use a five-tone, or *pentatonic,* scale. Whichever scale is used, each note has a characteristic frequency, and the notes span a frequency range called an *octave.*

Sound Quality of Instruments

The difference in sound quality among different instruments comes from their structural differences. All instruments produce sound by vibrating. But instruments vary in the part that vibrates and in the way that the vibrations are made. There are three main families of instruments: string instruments, wind instruments, and percussion instruments.

sound quality the result of the blending of several pitches through interference

✓ **Reading Check** How do musical instruments differ in how they produce sound? (*See the Appendix for answers to Reading Checks.*)

String Instruments

Violins, guitars, and banjos are examples of string instruments. They make sound when their strings vibrate after being plucked or bowed. **Figure 2** shows how two different string instruments produce sounds.

Figure 2   String Instruments

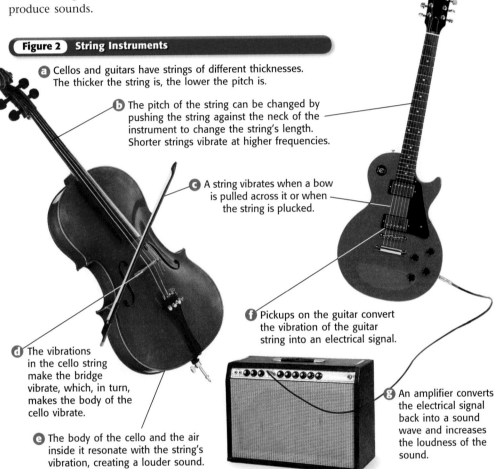

ⓐ Cellos and guitars have strings of different thicknesses. The thicker the string is, the lower the pitch is.

ⓑ The pitch of the string can be changed by pushing the string against the neck of the instrument to change the string's length. Shorter strings vibrate at higher frequencies.

ⓒ A string vibrates when a bow is pulled across it or when the string is plucked.

ⓓ The vibrations in the cello string make the bridge vibrate, which, in turn, makes the body of the cello vibrate.

ⓔ The body of the cello and the air inside it resonate with the string's vibration, creating a louder sound.

ⓕ Pickups on the guitar convert the vibration of the guitar string into an electrical signal.

ⓖ An amplifier converts the electrical signal back into a sound wave and increases the loudness of the sound.

Answer to Reading Check
Musical instruments differ in the part of the instrument that vibrates and in the way that the vibrations are made.

📖 **READING STRATEGY** — GENERAL

Prediction Guide Before students read these two pages, ask them to name the three major families of musical instruments and to give as many examples of each family as they can.
LS Logical

Demonstration — GENERAL

Families of Instruments Some of your students may be in the school orchestra or band. Ask them to bring in their musical instruments. Let them explain and demonstrate how their instruments work. Have students identify which family of instruments theirs are from and the range of frequency and pitch their instruments can make. Then, let the band or orchestra members play for the class!
LS Auditory

Homework — GENERAL

The Sound of Music Have students write in their **science journal** about a sound or some music that is important to them. The sound or music may have changed their perspective about something, or maybe it changed their attitude or mood. Maybe it is just a favorite song. Ask them to explain why they think music can produce a change in a person's mood. LS Intrapersonal

Musical Instruments and Sound Quality Have students name the three classes of musical instruments named in this section. Write each one on the board. Then, have them brainstorm as many examples of each one as they can think of. Write these in a second area on the board underneath each category. Then, have them explain what all instruments of each category have in common in how they produce sound and what determines their particular sound quality. **LS Logical**

1. What is sound quality? (the result of several pitches blending together through interference)

2. How does noise differ from music? (Music contains repeating melodic patterns. Noise has a complex sound wave with no pattern.)

Instrument Diagrams Have students draw and label members of each instrument family. Instruct them to label the part that vibrates and show where or how pitch can be changed. **LS Verbal** | English Language Learners

---

Figure 3 Wind Instruments

a A trumpet player's lips vibrate when the player blows into a trumpet.

b The reed vibrates back and forth when a musician blows into a clarinet.

c Standing waves are formed in the air columns of the instruments. The pitch of the instrument depends in part on the length of the air column. The longer the column is, the lower the pitch is.

d The length of the air column in a trumpet is changed by pushing the valves.

e The length of the air column in a clarinet is changed by closing or opening the finger holes.

Wind Instruments

A wind instrument produces sound when a vibration is created at one end of its air column. The vibration causes standing waves inside the air column. Pitch is changed by changing the length of the air column. Wind instruments are sometimes divided into two groups—woodwinds and brass. Examples of woodwinds are saxophones, oboes, and recorders. French horns, trombones, and tubas are brass instruments. A brass instrument and a woodwind instrument are shown in **Figure 3.**

Percussion Instruments

Drums, bells, and cymbals are percussion instruments. They make sound when struck. Instruments of different sizes are used to get different pitches. Usually, the larger the instrument is, the lower the pitch is. The drums and cymbals in a trap set, shown in **Figure 4,** are percussion instruments.

Figure 4 Percussion Instruments

The skins of the drums vibrate when struck with drumsticks.

Cymbals vibrate when struck together or when struck with drumsticks.

Each drum in the set is a different size. The larger the drum is, the lower the pitch is.

Cultural Awareness ── GENERAL

Use of Drums Drums, the most common percussion instruments, have been around since at least 6000 BCE. Drums are found in almost every culture and have been used for a number of purposes, including music. In some African cultures, drums were used to transmit messages over many miles. In Europe, infantry regiments once used snare drums to transmit coded orders to soldiers.

Music or Noise?

Most of the sounds we hear are noises. The sound of a truck roaring down the highway, the slam of a door, and the jingle of keys falling to the floor are all noises. **Noise** can be described as any sound, especially a nonmusical sound, that is a random mix of frequencies (or pitches). **Figure 5** shows on an oscilloscope the difference between a musical sound and noise.

noise a sound that consists of a random mix of frequencies

✓ **Reading Check** What is the difference between music and noise?

French horn

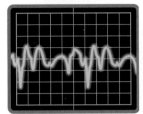

A sharp clap

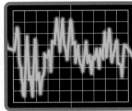

Figure 5 *A note from a French horn produces a sound wave with a repeating pattern, but noise from a clap produces complex sound waves with no regular pattern.*

SECTION Review

Summary

- Different instruments have different sound qualities.
- Sound quality results from the blending through interference of the fundamental and several overtones.
- The three families of instruments are string, wind, and percussion instruments.
- Noise is a sound consisting of a random mix of frequencies.

Using Key Terms

1. Use each of the following terms in a separate sentence: *sound quality* and *noise.*

Understanding Key Ideas

2. What interaction of sound waves determines sound quality?
   a. reflection   **c.** pitch
   b. diffraction   **d.** interference

3. Why do different instruments have different sound qualities?

Critical Thinking

4. **Making Comparisons** What do string instruments and wind instruments have in common in how they produce sound?

5. **Identifying Bias** Someone says that the music you are listening to is "just noise." Does the person mean that the music is a random mix of frequencies? Explain your answer.

Interpreting Graphics

6. Look at the oscilloscope screen below. Do you think the sound represented by the wave on the screen is noise or music? Explain your answer.

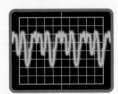

For a variety of links related to this chapter, go to www.scilinks.org

Topic: Sound Quality
SciLinks code: HSM1427

Answers to Section Review

1. Sample answer: Each instrument has a different sound quality based on how it produces sound. Static from the TV or radio is an example of noise.

2. d

3. A musical instrument's particular shape and way of producing sound give it its unique sound quality.

4. String instruments and wind instruments both make standing waves, string instruments along the string and wind instruments inside the air column, to produce sound.

5. probably not; People use the word *noise* in common speech to mean any unpleasant sound, which might be applied by someone to a particular kind of music even if that music is not merely a random mix of frequencies.

6. The sound is music. The oscilloscope shows an image of a wave with a repeating pattern.

Answer to Reading Check

Music consists of sound waves that have regular patterns, and noise consists of a random mix of frequencies.

CHAPTER RESOURCES

Chapter Resource File

- Section Quiz GENERAL
- Section Review GENERAL
- Vocabulary and Section Summary GENERAL
- Critical Thinking ADVANCED

Easy Listening

Teacher's Notes

Time Required

One or two 45-minute class periods

Lab Ratings

EASY ——————————→ HARD

Teacher Prep
Student Set-Up
Concept Level
Clean Up

MATERIALS

FOR EACH GROUP OF 3 TO 4 STUDENTS
• eraser, hard rubber (or tuning fork mallet)
• meterstick
• paper, graph
• tuning forks, different frequencies (4)

Form a Hypothesis

2. Sample answer: Most of the students in the class will hear mid-frequency sounds better. (Accept any testable hypothesis.)

Using Scientific Methods

Skills Practice Lab

Easy Listening

Pitch describes how low or high a sound is. A sound's pitch is related to its frequency—the number of waves per second. Frequency is measured in hertz (Hz), where 1 Hz equals 1 wave per second. Most humans can hear frequencies in the range from 20 Hz to 20,000 Hz. But not everyone detects all pitches equally well at all distances. In this activity, you will collect data to see how well you and your classmates hear different frequencies at different distances.

Ask a Question

1. Do most of the students in your classroom hear low-, mid-, or high-frequency sounds best?

Form a Hypothesis

2. Write a hypothesis that answers the question above. Explain your reasoning.

Test the Hypothesis

3. Choose one member of your group to be the sound maker. The others will be the listeners.

4. Copy the data table below onto another sheet of paper. Be sure to include a column for every listener in your group.

OBJECTIVES

Measure your classmates' ability to detect different pitches at different distances.

Graph the average class data.

Form a conclusion about how easily pitches of different frequencies are heard at different distances.

MATERIALS

• eraser, hard rubber
• meterstick
• paper, graph
• tuning forks, different frequencies (4)

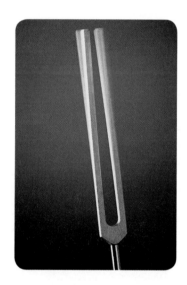

Data Collection Table				
	Distance (m)			
Frequency	Listener 1	Listener 2	Listener 3	Average
1 (____ Hz)				
2 (____ Hz)		DO NOT WRITE IN BOOK		
3 (____ Hz)				
4 (____ Hz)				

5. The sound maker will choose one of the tuning forks, and record the frequency of the tuning fork in the data table.

6. The listeners should stand 1 m from the sound maker with their backs turned.

Lab Notes

You may wish to use the classroom graph to have students practice interpreting a graph. For instance, you could ask students to try and pinpoint the distances at which other frequencies might be heard based on the graph.

CHAPTER RESOURCES

Chapter Resource File
• Datasheet for Chapter Lab
• Lab Notes and Answers

Technology

Classroom Videos
• Lab Video

LabBook

• The Speed of Sound
• Tuneful Tube

7 The sound maker will create a sound by striking the tip of the tuning fork gently with the eraser.

8 Listeners who hear the sound should take one step away from the sound maker. The listeners who do not hear the sound should stay where they are.

9 Repeat steps 7 and 8 until none of the listeners can hear the sound or the listeners reach the edge of the room.

10 Using the meterstick, the sound maker should measure the distance from his or her position to each of the listeners. All group members should record this data.

11 Repeat steps 5 through 10 with a tuning fork of a different frequency.

12 Continue until all four tuning forks have been tested.

Analyze the Results

1 **Organizing Data** Calculate the average distance for each frequency. Share your group's data with the rest of the class to make a data table for the whole class.

2 **Analyzing Data** Calculate the average distance for each frequency for the class.

3 **Constructing Graphs** Make a graph of the class results, plotting average distance (*y*-axis) versus frequency (*x*-axis).

Draw Conclusions

4 **Drawing Conclusions** Was everyone in the class able to hear all of frequencies equally? (Hint: Was the average distance for each frequency the same?)

5 **Evaluating Data** If the answer to question 4 is no, which frequency had the longest average distance? Which frequency had the shortest final distance?

6 **Analyzing Graphs** Based on your graph, do your results support your hypothesis? Explain your answer.

7 **Evaluating Methods** Do you think your class sample is large enough to confirm your hypothesis for all people of all ages? Explain your answer.

Terry Rakes
Elmwood Junior High
Rogers, Arkansas

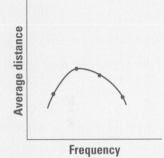

Chapter Review

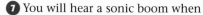

Assignment Guide

Section	Questions
1	11–12, 17
2	1, 4–5, 9–10, 13, 19–20
3	2, 6–8, 15–16
4	3, 18, 21
2 and 3	14

ANSWERS

Using Key Terms

1. loudness
2. echoes
3. sound quality

Understanding Key Ideas

4. a
5. d
6. c
7. c
8. d
9. b
10. If a fish is moving away from the whale, the echo off the fish that the whale hears will have a lower pitch than the original sound. If the fish is moving toward the whale, the echo off the fish heard by the whale will have a higher pitch than the original sound.
11. The back-and-forth motions of vibrations of an object cause compressions and rarefactions in the air around it, which is carried outward as sound waves.

USING KEY TERMS

Complete each of the following sentences by choosing the correct term from the word bank.

loudness   echoes
pitch   noise
sound quality

1. The _____ of a sound wave depends on its amplitude.

2. Reflected sound waves are called _____.

3. Two different instruments playing the same note sound different because of _____.

UNDERSTANDING KEY IDEAS

Multiple Choice

4. If a fire engine is traveling toward you, the Doppler effect will cause the siren to sound
   a. higher. c. louder.
   b. lower. d. softer.

5. Sound travels fastest through
   a. a vacuum. c. air.
   b. sea water. d. glass.

6. If two sound waves interfere constructively, you will hear
   a. a high-pitched sound.
   b. a softer sound.
   c. a louder sound.
   d. no change in sound.

7. You will hear a sonic boom when
   a. an object breaks the sound barrier.
   b. an object travels at supersonic speeds.
   c. a shock wave reaches your ears.
   d. the speed of sound is 290 m/s.

8. Resonance can happen when an object vibrates at another object's
   a. resonant frequency.
   b. fundamental frequency.
   c. second overtone frequency.
   d. All of the above

9. A technological device that can be used to see sound waves is a(n)
   a. sonar. c. ultrasound.
   b. oscilloscope. d. amplifier.

Short Answer

10. Describe how the Doppler effect helps a beluga whale determine whether a fish is moving away from it or toward it.

11. How do vibrations cause sound waves?

12. Briefly describe what happens in the different parts of the ear.

Math Skills

13. A submarine that is not moving sends out a sonar sound wave traveling 1,500 m/s, which reflects off a boat back to the submarine. The sonar crew detects the reflected wave 6 s after it was sent out. How far away is the boat from the submarine?

12. In the outer ear, sound waves are funneled into the ear canal. In the middle ear, the hammer, anvil, and stirrup increase the size of the vibrations. In the inner ear, these vibrations are changed into electrical signals for the brain to interpret.

13. 1,500 m/s × 6 s = 9,000 m
    9,000 m ÷ 2 = 4,500 m

CRITICAL THINKING

14 Concept Mapping Use the following terms to create a concept map: *sound waves, pitch, loudness, decibels, frequency, amplitude, oscilloscope, hertz,* and *interference.*

15 Analyzing Processes An *anechoic chamber* is a room where there is almost no reflection of sound waves. Anechoic chambers are often used to test sound equipment, such as stereos. The walls of such chambers are usually covered with foam triangles. Explain why this design eliminates echoes in the room.

16 Applying Concepts Would the pilot of an airplane breaking the sound barrier hear a sonic boom? Explain why or why not.

17 Forming Hypotheses After working in a factory for a month, a man you know complains about a ringing in his ears. What might be wrong with him? What do you think may have caused his problem? What can you suggest to him to prevent further hearing loss?

INTERPRETING GRAPHICS

Use the oscilloscope screens below to answer the questions that follow:

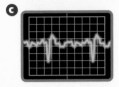

18 Which sound is noise?

19 Which represents the softest sound?

20 Which represents the sound with the lowest pitch?

21 Which two sounds were produced by the same instrument?

Critical Thinking

14. An answer to this exercise can be found at the end of this book.

15. This design eliminates echoes because the surfaces of the walls absorb most sound waves instead of reflecting them.

16. no; A sonic boom is only audible to an observer behind the plane along the shock wave. Once the airplane breaks the sound barrier, the plane outruns the sonic boom, so the pilot does not hear it.

17. The worker may be suffering from tinnitus, caused by long exposure to loud sounds. Further hearing loss could be prevented by wearing hearing protection while on the job.

Interpreting Graphics

18. d

19. a

20. c

21. b, c

CHAPTER RESOURCES

Chapter Resource File

- Chapter Review **GENERAL**
- Chapter Test A **GENERAL**
- Chapter Test B **ADVANCED**
- Chapter Test C **SPECIAL NEEDS**
- Vocabulary Activity **GENERAL**

Workbooks

Study Guide
- Assessment resources are also available in Spanish.

Standardized Test Preparation

Teacher's Note

To provide practice under more realistic testing conditions, give students 20 minutes to answer all of the questions in this Standardized Test Preparation.

MISCONCEPTION ALERT

Answers to the standardized test preparation can help you identify student misconceptions and misunderstandings.

READING

Passage 1

1. D
2. G
3. A

TEST DOCTOR

Question 2: The key to this question is zeroing in on the one most defining aspect of booming sands among the descriptors shown. Although they are found in Asia, this is not a defining aspect of booming sands: they are found elsewhere. Booming sands can be caused by slippage of sand over sand dunes, but "slippery" is not the best way to describe them. Booming sands were described by Marco Polo, but this is, again, not a defining aspect of booming sands. The best answer choice is G, "noisy," because this is the one quality referred to most throughout the passage.

Standardized Test Preparation

READING

Read each of the passages below. Then, answer the questions that follow each passage.

Passage 1 Centuries ago, Marco Polo wrote about the booming sand dunes of the Asian desert. He wrote that the booming sands filled the air with the sounds of music, drums, and weapons of war. Booming sands are most often found in the middle of large deserts. They have been discovered all over the world, including the United States. Booming sands make loud, low-pitched sounds when the top layers of sand slip over the layers below, producing vibrations. The sounds have been compared to foghorns, cannon fire, and moaning. The sounds can last from a few seconds to 15 min and can be heard more than 10 km away!

1. Which is a fact in this passage?
   - **A** Marco Polo loved traveling.
   - **B** Booming sands always sound like moaning people.
   - **C** Booming sands are the most interesting thing in Asia.
   - **D** Some booming sands are found in the United States.

2. Which of the following phrases **best** describes booming sands?
   - **F** found in Asia
   - **G** noisy
   - **H** slippery
   - **I** discovered by Marco Polo

3. What causes booming sands?
   - **A** vibrations caused by top layers of sand slipping over layers below
   - **B** battles in the desert
   - **C** animals that live beneath sand dunes
   - **D** There is not enough information to determine the answer.

Passage 2 People who work in the field of architectural acoustics are concerned with controlling sound that travels in a closed space. Their goal is to make rooms and buildings quiet yet suitable for people to enjoy talking and listening to music. One major factor that affects the acoustical quality of a room is the way the room reflects sound waves. Sound waves bounce off surfaces such as doors, ceilings, and walls. Using materials that absorb sound reduces the reflection of sound waves. Materials that have small pockets of air that can trap the sound vibrations and keep them from reflecting are the most sound absorbent. Sound-absorbing floor and ceiling tiles, curtains, and upholstered furniture all help to control the reflection of sound waves.

1. The field of architectural acoustics is concerned with which of the following?
   - **A** making buildings earthquake safe
   - **B** controlling sound in closed spaces
   - **C** designing sound-absorbing materials
   - **D** making buildings as quiet as possible

2. Which of the following is a major factor in the acoustical quality of a room?
   - **F** the size of the room
   - **G** the furnishings in the room
   - **H** the walls of the room
   - **I** the noise level in the room

3. Which of the following materials is **most** likely to absorb sounds the best?
   - **A** materials that have small pockets of air
   - **B** surfaces such as doors, ceilings, and walls
   - **C** materials that keep the room as quiet as possible
   - **D** furniture that is made of wood

Passage 2

1. B
2. H
3. A

TEST DOCTOR

Question 3: Students may be tempted to select answer C, but it is not an actual description of the types of materials that are likely to absorb sound. Furniture is mentioned in the passage as a minor factor in room acoustics, but the passage names materials that have small pockets of air as the best sound-absorbing material. Answer A is therefore correct.

Use the pictures of standing waves below to answer the questions that follow.

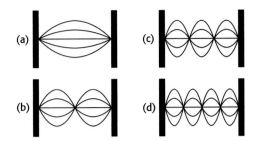

(a) (b) (c) (d)

1. Which of the standing waves has the lowest frequency?

A a
B b
C c
D d

2. Which of the standing waves has the highest frequency?

F a
G b
H c
I d

3. Which of the standing waves represents the first overtone?

A a
B b
C c
D d

4. In which of the following pairs of standing waves is the frequency of the second wave twice the frequency of the first?

F a, b
G a, c
H b, c
I c, d

Read each question below, and choose the best answer.

1. The speed of sound in copper is 3,560 m/s. Which is another way to express this measure?

A 356×10^2 m/s
B 0.356×10^3 m/s
C 3.56×10^3 m/s
D 3.56×10^4 m/s

2. The speed of sound in sea water is 1,522 m/s. How far can a sound wave travel underwater in 10 s?

F 152.2 m
G 1,522 m
H 15,220 m
I 152,220 m

3. Claire likes to go swimming after work. She warms up for 120 s before she begins swimming, and it takes her an average of 55 s to swim one lap. Which equation could be used to find w, the number of seconds it takes for Claire to warm up and swim 15 laps?

A $w = (15 \times 120) + 55$
B $w = (15 \times 55) + 120$
C $w = 120 + 55 + 15$
D $w = (15 \times 55) \times 120$

4. The Vasquez family went bowling. They rented 6 pairs of shoes for $3 a pair and bowled for 2 h at a rate of $8.80/h. Which is the best estimate of the total cost of the shoes and bowling?

F $24
G $30
H $36
I $45

Standardized Test Preparation

1. A
2. I
3. B
4. F

➕ **TEST DOCTOR**

Question 3: The first overtone should not be confused with the first harmonic, which is the same as the fundamental. The first overtone is the standing wave with twice the frequency (and half the wavelength) of the fundamental.

1. C
2. H
3. B
4. H

➕ **TEST DOCTOR**

Question 3: If students recognize that the order of operations for addition is not important, they will recognize B as the correct answer. If they do not, they may think that the fact that the 120 s of warmup is mentioned first in the passage means they should look for the 120 s to come first in the equation, and they may be tempted to choose C.

Science in Action

Science in Action

Science, Technology, and Society

Background

Computed axial tomography (CAT scanning), which has been used for many years in medicine, is now being used by paleontologists to study the internal structure of fossils. CAT scanning can provide interior views of a fossil without touching the fossil's surface. If a paleontologist needs to reconstruct an entire skull, a series of two-dimensional "slice" shots are taken and combined through computer imaging to produce a three-dimensional image of the skull—inside and out!

Science Fiction

Background

As a Caldecott Award winner, National Book nominee, and Nebula finalist, Jane Yolen knows how to write successful stories. Her work spans a wide range of topics—from imaginative alphabet books for the youngest reader to serious novels for the young-adult and adult audience. The inspiration for much of Yolen's work comes from folktales and stories with rich histories.

Scientific Discoveries

Jurassic Bark

Imagine you suddenly hear a loud honking sound, such as a trombone or a tuba. "Must be band tryouts," you think. You turn to find the noise and find yourself face to face with a 10 m long, 2,800 kg dinosaur with a huge tubular crest on its snout. Do you run? No—your musical friend, *Parasaurolophus,* is a vegetarian. In 1995, an almost-complete fossil skull of an adult *Parasaurolophus* was found in New Mexico. Scientists studied the noise-making qualities of *Parasaurolophus*'s crest and found that it contained many internal tubes and chambers.

Math ACTiViTY

Imagine that a standing wave with a frequency of 80 Hz is made inside the crest of a *Parasaurolophus*. What would be the frequency of the first overtone of this standing wave? the second? the third?

Science Fiction

"Ear" by Jane Yolen

Jily and her friends, Sanya and Feeny, live in a time not too far in the future. It is a time when everyone's hearing is damaged. People communicate using sign language—unless they put on their Ear. Then, the whole world is filled with sounds.

Jily and her friends visit a club called The Low Down. It is too quiet for Jily's tastes, and she wants to leave. But Sanya is dancing by herself, even though there is no music. When Jily finds Feeny, they notice some Earless kids their own age. Earless people never go to clubs, and Jily finds their presence offensive. But Feeny is intrigued.

Everyone is given an Ear at the age of 12 but has to give it up at the age of 30. Why would these kids want to go out without their Ears before the age of 30? Jily thinks the idea is ridiculous and doesn't stick around to find out the answer to such a question. But it is an answer that will change her life by the end of the next day.

Language Arts ACTiViTY

WRITING SKILL Read "Ear," by Jane Yolen, in the *Holt Anthology of Science Fiction.* Write a one-page report that discusses how the story made you think about the importance of hearing in your everyday life.

Answer to Math Activity

The first overtone would be twice the frequency of the fundamental, so the frequency would be 80 Hz × 2 = 160 Hz. The second overtone would be 80 Hz × 3 = 240 Hz. The third overtone would be 80 Hz × 4 = 320 Hz.

Answer to Language Arts Activity

Answers may vary.

Adam Dudley

Sound Engineer Adam Dudley uses the science of sound waves every day at his job. He is the audio supervisor for the Performing Arts Center of the University of Texas at Austin. Dudley oversees sound design and technical support for campus performance spaces, including an auditorium that seats over 3,000 people.

To stage a successful concert, Dudley takes many factors into account. The size and shape of the room help determine how many speakers to use and where to place them. It is a challenge to make sure people seated in the back row can hear well enough and also to make sure that the people up front aren't going deaf from the high volume.

Adam Dudley loves his job—he enjoys working with people and technology and prefers not to wear a coat and tie. Although he is invisible to the audience, his work backstage is as crucial as the musicians and actors on stage to the success of the events.

Social Studies

ACTIVITY

Research the ways in which concert halls were designed before the use of electricity for amplification. Make a model or diorama, and present it to the class, explaining the acoustical factors involved in the design.

To learn more about these Science in Action topics, visit **go.hrw.com** and type in the keyword **HP5SNDF**.

Current Science

Check out Current Science® articles related to this chapter by visiting go.hrw.com. Just type in the keyword HP5CS21.

Careers

Background

The physics of sound is very important in the field of sound engineering. For instance, speakers placed in different locations in a concert hall must be placed in such a way that destructive interference does not create dead spots. Also, sometimes extra speakers are placed in the balcony to reinforce the sound coming from the speakers near the stage. But because of the difference in speed between the signal coming through the wire and the sound traveling through air, people in the balcony may hear a delay between the sound coming from the balcony speakers and that coming from speakers near the stage. This problem requires a correction in the signal to the balcony speakers, delaying it by a fraction of a second to compensate.

Answer to Social Studies Activity
Answers may vary.

3

The Nature of Light
Chapter Planning Guide

Compression guide:
To shorten instruction because of time limitations, omit the Chapter Lab.

OBJECTIVES	LABS, DEMONSTRATIONS, AND ACTIVITIES	TECHNOLOGY RESOURCES
PACING • 90 min pp. 60–65 **Chapter Opener**	SE **Start-up Activity,** p. 61 GENERAL	OSP **Parent Letter** GENERAL CD **Student Edition on CD-ROM** CD **Guided Reading Audio CD** ■ TR **Chapter Starter Transparency*** VID **Brain Food Video Quiz**
Section 1 What Is Light? • Describe light as an electromagnetic wave. • Calculate distances traveled by light by using the speed of light. • Explain why light from the sun is important.	TE **Demonstration** Glowing Green, p. 62 GENERAL SE **Science in Action** Math, Social Studies, and Language Arts Activities, pp. 94–95 GENERAL	CRF **Lesson Plans*** TR **Bellringer Transparency*** TR **Electromagnetic Wave*** CD **Science Tutor**
PACING • 45 min pp. 66–73 **Section 2 The Electromagnetic Spectrum** • Identify how electromagnetic waves differ. • Describe some uses for radio waves and microwaves. • List examples of how infrared waves and visible light are important in your life. • Explain how ultraviolet light, X rays, and gamma rays can be both helpful and harmful.	TE **Connection Activities** History, p. 67; Real Life, p. 68; Life Science, p. 69; Math, p. 71 GENERAL SE **School-to-Home Activity** Making a Rainbow, p. 70 GENERAL TE **Activity** Rainbow Research, p. 70 ADVANCED TE **Activity** Blocking UV Light, p. 71 BASIC SE **Connection to Astronomy** Gamma Ray Spectrometer, p. 72 GENERAL	CRF **Lesson Plans*** TR **Bellringer Transparency*** TR **The Electromagnetic Spectrum*** TR **LINK TO EARTH SCIENCE** The H-R Diagram: A and B* CRF **SciLinks Activity*** GENERAL CD **Science Tutor**
PACING • 45 min pp. 74–81 **Section 3 Interactions of Light Waves** • Describe how reflection allows you to see things. • Describe absorption and scattering. • Explain how refraction can create optical illusions and separate white light into colors. • Explain the relationship between diffraction and wavelength. • Compare constructive and destructive interference of light.	TE **Group Activity** Making a Periscope, p. 74 GENERAL TE **Activity** The Law of Reflection, p. 75 BASIC SE **Connection to Astronomy** Moonlight?, p. 76 GENERAL TE **Activity** Why Is the Sky Blue?, p. 76 ADVANCED SE **Quick Lab** Scattering Milk, p. 77 GENERAL TE **Connection Activity** Real World, p. 77 GENERAL TE **Connection Activity** Life Science, p. 78 ADVANCED SE **Quick Lab** Refraction Rainbow, p. 79 GENERAL TE **Demonstration** Diffraction, p. 79 GENERAL TE **Connection Activity** Real World, p. 80 GENERAL	CRF **Lesson Plans*** TR **Bellringer Transparency*** TR **The Law of Reflection/Regular Reflection Versus Diffuse Reflection*** SE **Internet Activity,** p. 80 GENERAL CD **Science Tutor**
PACING • 90 min pp. 82–87 **Section 4 Light and Color** • Name and describe the three ways light interacts with matter. • Explain how the color of an object is determined. • Explain why mixing colors of light is called *color addition*. • Describe why mixing colors of pigments is called *color subtraction*.	TE **Demonstration** Adding Colors, p. 82 GENERAL TE **Group Activity** Colorblindness, p. 83 GENERAL TE **Connection Activity** Earth Science, p. 84 ADVANCED SE **School-to-Home Activity** Television Colors, p. 85 GENERAL TE **Demonstration** Adding Colors, Part 2, p. 85 GENERAL TE **Connection Activity** Art, p. 85 ADVANCED SE **Quick Lab** Rose-Colored Glasses?, p. 86 GENERAL SE **Skills Practice Labs** p. 88, p. 130, p. 131 GENERAL LB **Long-Term Projects & Research Ideas** The Image of the Future* ADVANCED	CRF **Lesson Plans*** TR **Bellringer Transparency*** TR **Color Addition*/Color Subtraction*** CD **Interactive Explorations CD-ROM** In the Spotlight GENERAL VID **Lab Videos for Physical Science** CD **Science Tutor**

PACING • 90 min

CHAPTER REVIEW, ASSESSMENT, AND STANDARDIZED TEST PREPARATION

CRF **Vocabulary Activity*** GENERAL
SE **Chapter Review,** pp. 90–91 GENERAL
CRF **Chapter Review*** ■ GENERAL
CRF **Chapter Tests A*** ■ GENERAL, **B*** ADVANCED, **C*** SPECIAL NEEDS
SE **Standardized Test Preparation,** pp. 92–93 GENERAL
CRF **Standardized Test Preparation*** GENERAL
CRF **Performance-Based Assessment*** GENERAL
OSP **Test Generator** GENERAL
CRF **Test Item Listing*** GENERAL

Online and Technology Resources

Visit **go.hrw.com** for a variety of free resources related to this textbook. Enter the keyword **HP5LGT**.

Holt Online Learning

Students can access interactive problem-solving help and active visual concept development with the *Holt Science and Technology* Online Edition available at **www.hrw.com**.

 Guided Reading Audio CD
Also in Spanish

A direct reading of each chapter for auditory learners, reluctant readers, and Spanish-speaking students.

 Science Tutor CD-ROM

Excellent for remediation and test practice.

SKILLS DEVELOPMENT RESOURCES	SECTION REVIEW AND ASSESSMENT	CORRELATIONS
SE Pre-Reading Activity, p. 60 GENERAL **OSP** Science Puzzlers, Twisters & Teasers* GENERAL		National Science Education Standards SAI 1
CRF Directed Reading A* ■ BASIC, B* SPECIAL NEEDS **CRF** Vocabulary and Section Summary* ■ GENERAL **SE** Reading Strategy Brainstorming, p. 62 GENERAL **SE** Connection to Social Studies The Particle Model of Light, p. 63 GENERAL **TE** Inclusion Strategies, p. 63 **SE** Math Focus How Fast is Light?, p. 64 GENERAL	**SE** Reading Checks, pp. 63, 64 GENERAL **TE** Reteaching, p. 64 BASIC **TE** Quiz, p. 64 GENERAL **TE** Alternative Assessment, p. 64 GENERAL **SE** Section Review,* p. 65 ■ GENERAL **TE** Homework, p. 65 GENERAL **CRF** Section Quiz* ■ GENERAL	UCP 2, 3; PS 3a, 3f
CRF Directed Reading A* ■ BASIC, B* SPECIAL NEEDS **CRF** Vocabulary and Section Summary* ■ GENERAL **SE** Reading Strategy Mnemonics, p. 66 GENERAL **TE** Reading Strategy Prediction Guide, p. 67 GENERAL	**SE** Reading Checks, pp. 66, 68, 70, 71, 72 GENERAL **TE** Homework, p. 67 BASIC **TE** Reteaching, p. 72 BASIC **TE** Quiz, p. 72 GENERAL **TE** Alternative Assessment, p. 72 GENERAL **SE** Section Review,* p. 73 GENERAL **CRF** Section Quiz* ■ GENERAL	SPSP 1, 5; PS 3a, 3f
CRF Directed Reading A* ■ BASIC, B* SPECIAL NEEDS **CRF** Vocabulary and Section Summary* ■ GENERAL **SE** Reading Strategy Reading Organizer, p. 74 GENERAL **TE** Reading Strategy Prediction Guide, p. 75 GENERAL **CRF** Reinforcement Worksheet Light Interactions* BASIC **CRF** Critical Thinking Now You See It, Now You Don't ADVANCED	**SE** Reading Checks, pp. 74, 75, 76, 79, 80 GENERAL **TE** Homework, p. 75 ADVANCED **TE** Reteaching, p. 80 BASIC **TE** Quiz, p. 80 GENERAL **TE** Alternative Assessment, p. 80 GENERAL **SE** Section Review,* p. 81 ■ GENERAL **CRF** Section Quiz* ■ GENERAL	UCP 3; SAI 1; PS 3c
CRF Directed Reading A* ■ BASIC, B* SPECIAL NEEDS **CRF** Vocabulary and Section Summary* ■ GENERAL **SE** Reading Strategy Discussion, p. 82 GENERAL **TE** Inclusion Strategies, p. 83 **TE** Reading Strategy Prediction Guide, p. 84 GENERAL	**SE** Reading Checks, pp. 83, 84, 86 GENERAL **TE** Homework, p. 82 GENERAL **TE** Homework, p. 83 BASIC **TE** Reteaching, p. 86 BASIC **TE** Quiz, p. 86 GENERAL **TE** Alternative Assessment, p. 86 GENERAL **SE** Section Review,* p. 87 ■ GENERAL **CRF** Section Quiz* ■ GENERAL	PS 3a, 3c; *Chapter Lab:* SAI 1; PS 3c; *LabBook:* UCP 3; SAI 1; PS 3c

One-Stop Planner® CD-ROM

This CD-ROM package includes:
- Lab Materials QuickList Software
- Holt Calendar Planner
- Customizable Lesson Plans
- Printable Worksheets
- ExamView® Test Generator
- Interactive Teacher Edition
- Holt PuzzlePro® Resources
- Holt PowerPoint® Resources

SCLINKS®
NSTA
www.scilinks.org

Maintained by the **National Science Teachers Association.** See Chapter Enrichment pages for a complete list of topics.

Current Science®

Check out **Current Science** articles and activities by visiting the HRW Web site at **go.hrw.com.** Just type in the keyword **HP5CS22T.**

Classroom Videos

- **Lab Videos** demonstrate the chapter lab.
- **Brain Food Video Quizzes** help students review the chapter material.
- **CNN Videos** bring science into your students' daily life.

Visual Resources

CHAPTER STARTER TRANSPARENCY

Strange but True!

BELLRINGER TRANSPARENCIES

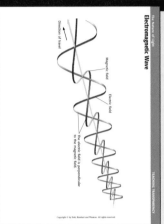

Section: What Is Light?
Some of the following questions have been debated for centuries. Record your responses to them in your **science journal.** What do you think light is? Is light made of matter, or is it purely energy? What is your reason for your answer? Can light travel through space?

Section: The Electromagnetic Spectrum
What are the weather conditions necessary to see a rainbow? Why do rainbows form? When else can you see a rainbow-like phenomenon?

Record your answers in your **science journal.**

TEACHING TRANSPARENCIES

Electromagnetic Wave

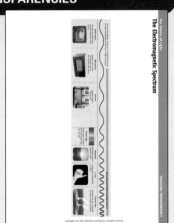

The Electromagnetic Spectrum

TEACHING TRANSPARENCIES

The Law of Reflection

Regular Reflection Versus Diffuse Reflection

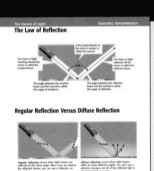

Color Addition

Color Subtraction

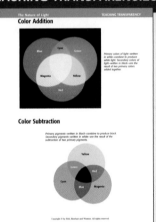

The H-R Diagram: A

The H-R Diagram

LINK TO EARTH SCIENCE

Chapter: Stars, Galaxies, and the Universe

CONCEPT MAPPING TRANSPARENCY

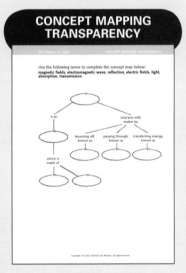

Use the following terms to complete the concept map below: **magnetic fields, electromagnetic wave, reflection, electric fields, light, absorption, transmission**

Planning Resources

LESSON PLANS

Lesson Plan — SAMPLE

Section: Waves

Pacing
Regular Schedule: with lab(s):2 days   without lab(s):if day
Block Schedule: with lab(s):1 1/2 days without lab(s) :if day

Objectives
1. Relate the seven properties of life to a living organism.
2. Describe seven themes that can help you to organize what you learn about biology.
3. Identify the tiny structures that make up all living organisms.
4. Differentiate between reproduction and heredity and between metabolism and homeostasis.

National Science Education Standards Covered

PARENT LETTER

SAMPLE

Dear Parent,

Your son's or daughter's science class will soon begin exploring the chapter entitled "The World of Physical Science." In this chapter, students will learn about how the scientific method applies to the world of physical science and the role of physical science in the world. By the end of the chapter, students should demonstrate a clear understanding of the chapter's main ideas and be able to discuss the following topics:

1. physical science as the study of energy and matter (Section 1)
2. the role of physical science in the world around them (Section 1)
3. careers that rely on physical science (Section 1)
4. the steps used in the scientific method (Section 2)
5. examples of technology (Section 2)
6. how the scientific method is used to answer questions and solve problems (Section 2)
7. how our knowledge of science changes over time (Section 2)
8. how models represent real objects or systems (Section 3)
9. examples of different ways models are used in science (Section 3)
10. the importance of the International System of Units (Section 4)
11. the appropriate units to use for particular measurements (Section 4)
12. how area and density are derived quantities (Section 4)

Questions to Ask Along the Way

You can help your son or daughter learn about these topics by asking interesting questions such as the following:
• What are some surprising careers that use physical science?
• What is a characteristic of a good hypothesis?
• When is it a good idea to use a model?
• Why do Americans measure things in terms of inches and yards and meters ?

ALSO IN SPANISH

TEST ITEM LISTING

TEST ITEM LISTING
The World of Science — SAMPLE

MULTIPLE CHOICE

Meeting Individual Needs

DIRECTED READING A

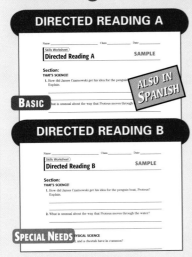

BASIC

DIRECTED READING B

SPECIAL NEEDS

VOCABULARY ACTIVITY

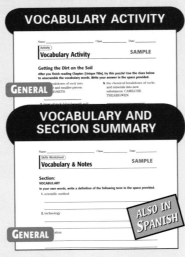

GENERAL

VOCABULARY AND SECTION SUMMARY

GENERAL

REINFORCEMENT

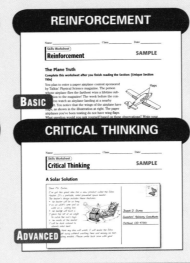

BASIC

CRITICAL THINKING

ADVANCED

SCILINKS ACTIVITY

GENERAL

SCIENCE PUZZLERS, TWISTERS & TEASERS

GENERAL

Labs and Activities

LONG-TERM PROJECTS & RESEARCH IDEAS

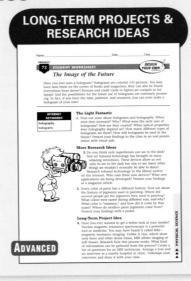

ADVANCED

DATASHEETS FOR QUICK LABS

DATASHEETS FOR CHAPTER LABS

DATASHEETS FOR LABBOOK

Review and Assessments

SECTION QUIZ

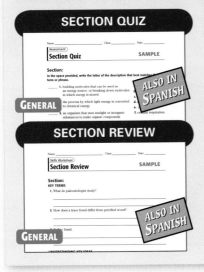

GENERAL

SECTION REVIEW

GENERAL

CHAPTER REVIEW

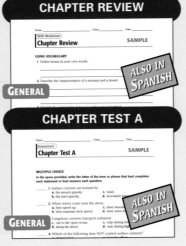

GENERAL

CHAPTER TEST A

GENERAL

CHAPTER TEST B

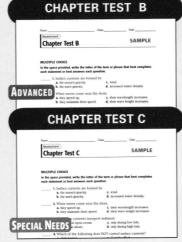

ADVANCED

CHAPTER TEST C

SPECIAL NEEDS

STANDARDIZED TEST PREPARATION

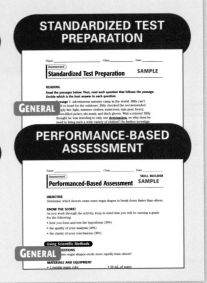

GENERAL

PERFORMANCE-BASED ASSESSMENT

GENERAL

This Chapter Enrichment provides relevant and interesting information to expand and enhance your presentation of the chapter material.

Section 1

What Is Light?
Theories of Light

- The nature of light has been debated for thousands of years. The argument about how light travels—as a wave or as a particle—began with the Greek mathematician and philosopher Pythagoras (c. 580–500 BCE). Pythagoras and his followers believed that light is emitted from a source in the form of tiny particles.

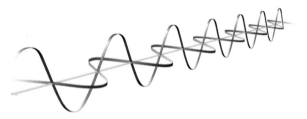

- However, Empedocles (c. 490–430 BCE), another Greek philosopher, taught that light travels from its source as waves.

- In the fifth century BCE, the Greek philosophers Socrates (c. 470–399 BCE) and Plato (c. 428–348 BCE) thought that the eyes emitted streamers, or filaments, and that sight occurred when these streamers made contact with objects.

- Even as late as the 1600s, René Descartes (1596–1650), the great French mathematician and philosopher, held beliefs similar to those of Socrates and Plato.

Is That a Fact!

◆ Galileo Galilei (1564–1642) once tried to measure the speed of light from a lantern from one hilltop to another. He soon realized that light traveled very fast.

Section 2

The Electromagnetic Spectrum
Maxwell's Equations

- In 1865, James Clerk Maxwell (1831–1879), a Scottish physicist, developed a theory stating that certain waves are propagated through space at the speed of light. In fact, his equations predicted that energy of the waves was equally divided between an electric field and a magnetic field. He called the waves electromagnetic waves.

Is That a Fact!

◆ The infrared portion of the spectrum was discovered by William Herschel (1738–1822), a famous British astronomer. In 1800, he was investigating the heat produced by certain waves located just below the red part of the visible spectrum. He named the waves *infrared*. *Infra* is a Latin word meaning "below."

◆ In 1895, Wilhelm Roentgen (1845–1923) serendipitously discovered X rays. While working with a barium ore, he discovered that the ore glowed when placed near a tube in which an electric current was passing. He experimented and found that the ore would glow even if placed behind substances that would block ordinary light.

◆ Roentgen didn't know the source of the radiation that caused the barium to glow, so he called the source X rays. He received the first Nobel Prize in physics in 1901 for his discovery of X rays.

- In 1886, Heinrich Hertz (1857–1894) was trying to prove Maxwell's equations experimentally. His experimentation was very fruitful: Not only did he prove Maxwell correct, but he also discovered radio waves. The unit for frequency was named in his honor.

Section 3

Interactions of Light Waves

Thomas Young

- Thomas Young (1773–1829) was a medical doctor born in Milverton, England. But Young had a variety of interests and wrote important papers in Egyptology and physics. In fact, Young aided in the translation of the Rosetta stone. He was also the first to propose the three-primary-color model for vision.

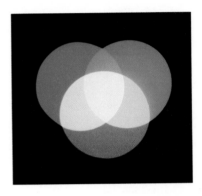

- In physics and the study of solid materials, the Young modulus (a measure of the strength and elasticity of a solid) is named after him for his work with solids.

- Young discovered the phenomenon of interference of light using an apparatus with a double slit. From his experiments, Young was able to measure the wavelengths of red and blue light. However, it was his revival of the wave theory of light that aided others in their search for the nature of light.

Section 4

Light and Color

Seeing Color

- In humans, light rays pass through the lens onto the retina. This back part of the eye contains two types of nerve cells, *rods* and *cones*, that respond to light energy.

- Cones, on the other hand, come in three types and are sensitive to different frequencies and intensities of light. One type is triggered by light energy at the blue end of the spectrum, the second responds to the red end of the spectrum, and the third type is stimulated by the middle of the spectrum, or the greens.

- Rods are most sensitive to movement and to changes in light and dark. Rods do not respond to different frequencies of light, so they do not perceive color.

- When light energy stimulates the cones, they send signals to the brain. The brain interprets these signals as colors depending on how many of each type of cone have been stimulated.

Is That a Fact!

- ◆ Experiments have shown that the human eye can detect a single photon of light. However, neural filters will only trigger a conscious response when between 5 and 10 photons arrive within a certain period of time. This threshold response keeps humans from seeing visual "noise" in low light!

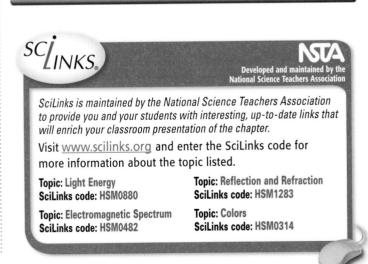

SCiLINKS

NSTA
Developed and maintained by the
National Science Teachers Association

SciLinks is maintained by the National Science Teachers Association to provide you and your students with interesting, up-to-date links that will enrich your classroom presentation of the chapter.

Visit www.scilinks.org and enter the SciLinks code for more information about the topic listed.

Topic: Light Energy
SciLinks code: HSM0880

Topic: Reflection and Refraction
SciLinks code: HSM1283

Topic: Electromagnetic Spectrum
SciLinks code: HSM0482

Topic: Colors
SciLinks code: HSM0314

Overview

Tell students that this chapter will help them learn about light as an electromagnetic wave. Students will learn about the electromagnetic spectrum and about interactions of light waves. Finally, students study the relationship between light and color.

Assessing Prior Knowledge

Students should be familiar with the following topics:

• properties of waves, including wavelength and frequency

Identifying Misconceptions

As students learn the material in this chapter, some of them may be confused about the divisions between the kinds of waves in the electromagnetic (EM) spectrum. Tell students that there are no strict dividing lines between the kinds of EM waves, and that many of the wavelength ranges overlap. The dividing wavelengths were set by people and are based on the ways certain EM waves were discovered and are used. Also, remind students that all EM waves are essentially the same; they only differ by wavelength and frequency.

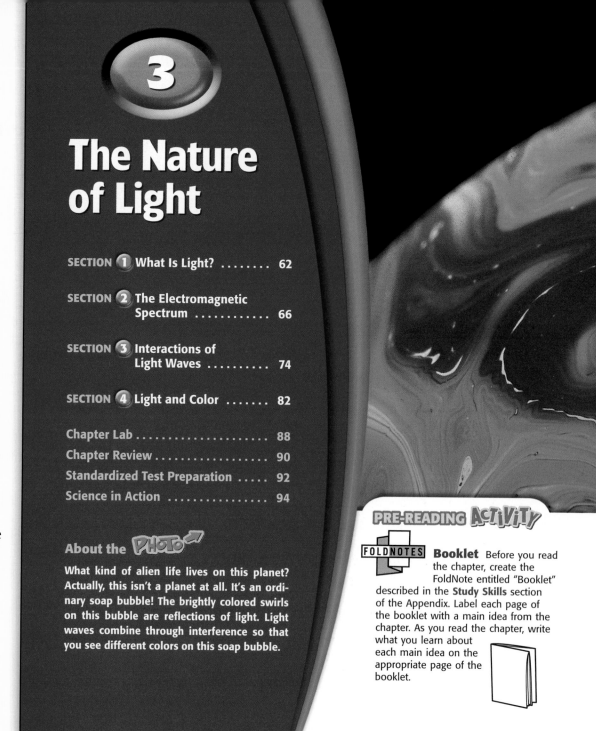

The Nature of Light

About the PHOTO

What kind of alien life lives on this planet? Actually, this isn't a planet at all. It's an ordinary soap bubble! The brightly colored swirls on this bubble are reflections of light. Light waves combine through interference so that you see different colors on this soap bubble.

PRE-READING ACTIVITY

FOLDNOTES **Booklet** Before you read the chapter, create the FoldNote entitled "Booklet" described in the **Study Skills** section of the Appendix. Label each page of the booklet with a main idea from the chapter. As you read the chapter, write what you learn about each main idea on the appropriate page of the booklet.

Standards Correlations

National Science Education Standards

The following codes indicate the National Science Education Standards that correlate to this chapter. The full text of the standards is at the front of the book.

Chapter Opener
SAI 1

Section 1 What Is Light?
UCP 2, 3; PS 3a, 3f

Section 2 The Electromagnetic Spectrum
SPSP 1, 5; PS 3a, 3f

Section 3 Interactions of Light Waves
UCP 3; SAI 1; PS 3c

Section 4 Light and Color
PS 3a, 3c; *LabBook:* UCP 3; SAI 1; PS 3c

Chapter Lab
SAI 1; PS 3c

Chapter Review
PS 3a, 3c

Science in Action
ST 2; SPSP 5; HNS 1, 3

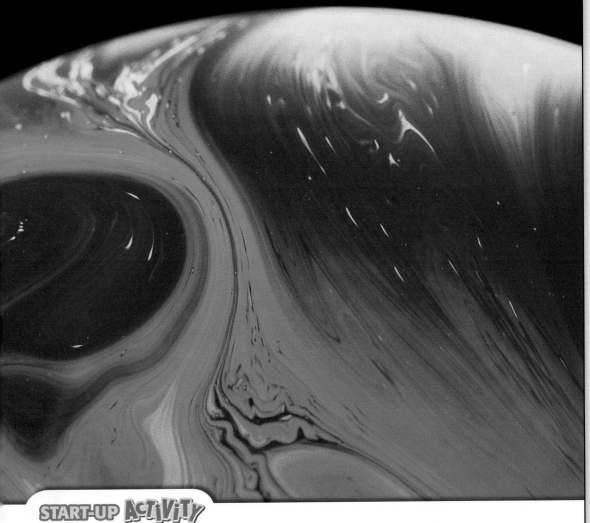

START-UP ACTIVITY

MATERIALS

FOR EACH GROUP
- diffraction grating
- light bulb, clear incandescent
- light bulb, fluorescent
- paper, construction, black
- paper-towel tube
- tape

Safety Caution: Students should avoid handling the hot bulbs.

Teacher's Notes: Students can make their own spectroscopes, or you can make them ahead of time. The instructions for making a spectroscope are:

1. Cut a narrow slit in the center of a piece of black construction paper. Tape the paper to one end of a paper-towel tube. The paper should cover the opening of the tube.

2. Look through the open end of the tube at an incandescent light bulb. If no light passes through the slit in the paper, make the slit a little wider.

3. Hold a diffraction grating against the open end of the tube. Look at the light bulb through the grating. Make sure the slit in the paper is vertical.

4. Rotate the diffraction grating until you see colors inside the tube to the left and right sides of the slit. Tape the diffraction grating to the tube in this position.

START-UP ACTIVITY

Colors of Light

Is white light really white? In this activity, you will use a spectroscope to answer that question.

Procedure

1. Your teacher will give you a **spectroscope** or instructions for making one.

2. Turn on an **incandescent light bulb.** Look at the light bulb through your spectroscope. Write a description of what you see.

3. Repeat step 2, looking at a **fluorescent light.** Again, describe what you see.

Analysis

1. Compare what you saw with the incandescent light bulb with what you saw with the fluorescent light bulb.

2. Both kinds of bulbs produce white light. What did you learn about white light by using the spectroscope?

3. Light from a flame is yellowish but is similar to white light. What do you think you would see if you used a spectroscope to look at light from a flame?

Answers to Start-Up Activity

1. Incandescent light: Students should see a continuous spectrum of colors on the sides of the tube. All colors should have the same brightness. Fluorescent light: Students should see a spectrum of colors. However, there will be bright bands and faint bands within the spectrum.

2. White light is made up of different colors of light.

3. Students should expect to see a spectrum of colors.

Chapter Starter Transparency
Use this transparency to help students begin thinking about the nature of light.

CHAPTER RESOURCES

Technology

Transparencies
- Chapter Starter Transparency

`READING SKILLS`

Student Edition on CD-ROM

Guided Reading Audio CD
- English or Spanish

Classroom Videos
- Brain Food Video Quiz

Workbooks

Science Puzzlers, Twisters & Teasers
- The Nature of Light `GENERAL`

Focus

Overview

Students will learn what electro-magnetic waves (EM waves) are, how they are produced, and how they differ from other waves. Students will also learn about the speed of light and light energy from the sun.

 **Bellringer**

Ask students to write answers to the following questions: "What do you think light is? Is light made of matter? Can light travel through space? Explain your answer."

Motivate

Demonstration — GENERAL

Glowing Green Using tongs to hold one end of a small piece of copper wire, place the other end into the flame of a Bunsen burner. (Any source of thermal energy, such as a lighter, will work.) A green, luminous glow will be produced. Ask students to explain the source of the green glow. Guide the discussion to help students realize that atoms in the copper wire emit a green light when thermal energy is added. **LS Visual/Verbal**

READING WARM-UP

Objectives

● Describe light as an electromagnetic wave.

● Calculate distances traveled by light by using the speed of light.

● Explain why light from the sun is important.

Terms to Learn

electromagnetic wave
radiation

READING STRATEGY

Brainstorming The key idea of this section is light. Brainstorm words and phrases related to light.

electromagnetic wave a wave that consists of electric and magnetic fields that vibrate at right angles to each other

---

What Is Light?

You can see light. It's everywhere! Light comes from the sun and from other sources, such as light bulbs. But what exactly is light?

Scientists are still studying light to learn more about it. A lot has already been discovered about light, as you will soon find out. Read on, and be enlightened!

Light: An Electromagnetic Wave

Light is a type of energy that travels as a wave. But light is different from other kinds of waves. Other kinds of waves, like sound waves and water waves, must travel through matter. Light does not require matter through which to travel. Light is an electromagnetic wave (EM wave). An **electromagnetic wave** is a wave that can travel through empty space or matter and consists of changing electric and magnetic fields.

Fields exist around certain objects and can exert a force on another object without touching that object. For example, Earth is a source of a gravitational field. This field pulls you and all things toward Earth. But keep in mind that this field, like all fields, is not made of matter.

Figure 1 shows a diagram of an electromagnetic wave. Notice that the electric and magnetic fields are at right angles—or are *perpendicular*—to each other. These fields are also perpendicular to the direction of the wave motion.

Figure 1 *Electromagnetic waves are made of vibrating electric and magnetic fields.*

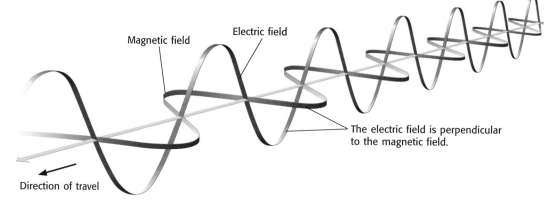

Magnetic field

Electric field

The electric field is perpendicular to the magnetic field.

Direction of travel

CHAPTER RESOURCES

Chapter Resource File

- Lesson Plan
- Directed Reading A **BASIC**
- Directed Reading B **SPECIAL NEEDS**

Technology

Transparencies
- Bellringer
- Electromagnetic Wave

Is That a Fact!

Galileo was perhaps the first scientist to suggest a method for measuring the speed of light. In 1676, the Danish astronomer Ole Roemer (1644–1710) was the first to demonstrate that the speed of light is finite by observing the eclipses of Jupiter's satellites.

Figure 2 *The hair on the girl's head stands up because of an electric field and the iron filings form arcs around the magnet because of a magnetic field.*

Electric and Magnetic Fields

Electromagnetic waves are changing electric and magnetic fields. But what are electric and magnetic fields? An *electric field* surrounds every charged object. The electric field around a charged object pulls oppositely charged objects toward it and repels like-charged objects. You can see the effect of electric fields whenever you see objects stuck together by static electricity. **Figure 2** shows another effect of an electric field.

A *magnetic field* surrounds every magnet. Because of magnetic fields, paper clips and iron filings are pulled toward magnets. You can feel the effect of magnetic fields when you hold two magnets close together. The iron filings around the magnet in **Figure 2** form arcs in the presence of the magnet's magnetic field.

☑ **Reading Check** Where can electric fields be found? (*See the Appendix for answers to Reading Checks.*)

How EM Waves Are Produced

An EM wave can be produced by the vibration of an electrically charged particle. When the particle vibrates, or moves back and forth, the electric field around it also vibrates. When the electric field starts vibrating, a vibrating magnetic field is created. The vibration of an electric field and a magnetic field together produces an EM wave that carries energy released by the original vibration of the particle. The transfer of energy as electromagnetic waves is called **radiation.**

CONNECTION TO Social Studies

WRITING SKILL **The Particle Model of Light**
Thinking of light as being an electromagnetic wave can explain many properties of light. But some properties of light can be explained only by using a particle model of light. In the particle model of light, light is thought of as a stream of particles called *photons*. Research the history of the particle model of light. Write a one-page paper on what you learn.

radiation transfer of energy as electromagnetic waves

Solar System Model Organize students into small groups, and have each group build a three-dimensional model of the solar system. Then, have students use their models to explain why so little of the sun's energy reaches Earth. LS **Kinesthetic/Interpersonal** Co-op Learning

Quiz ———————— GENERAL

1. The circumference of Earth is 40,100 km. How many times can light travel around Earth in 1 s? (about 7 times)

2. Where can magnetic fields be found? (Magnetic fields surround every magnet. Also acceptable: Magnetic fields are found in electromagnetic waves.)

3. What is radiation? (Radiation is the transfer of energy as electromagnetic waves.)

Alternate Assessment ——— GENERAL

Concept Mapping Have students make a concept map about light. The concept maps should illustrate the properties of light as an electromagnetic wave. LS **Visual/Verbal**

Figure 3 *Thunder and lightning are produced at the same time. But you usually see lightning before you hear thunder, because light travels much faster than sound.*

The Speed of Light

Scientists have yet to discover anything that travels faster than light. In the near vacuum of space, the speed of light is about 300,000,000 m/s, or 300,000 km/s. Light travels slightly slower in air, glass, and other types of matter. (Keep in mind that even though electromagnetic waves do not need to travel through matter, they can travel through many substances.)

Believe it or not, light can travel about 880,000 times faster than sound! This fact explains the phenomenon described in **Figure 3.** If you could run at the speed of light, you could travel around Earth 7.5 times in 1 s.

✓ **Reading Check** How does the speed of light compare with the speed of sound?

 **MATH FOCUS**

How Fast Is Light? The distance from Earth to the moon is 384,000 km. Calculate the time it takes for light to travel that distance.

Step 1: Write the equation for speed.

$$speed = \frac{distance}{time}$$

Step 2: Rearrange the equation by multiplying by time and dividing by speed.

$$time = \frac{distance}{speed}$$

Step 3: Replace *distance* and *speed* with the values given in the problem, and solve.

$$time = \frac{384,000 \text{ km}}{300,000 \text{ km/s}}$$

$$time = 1.28 \text{ s}$$

Now It's Your Turn

1. The distance from the sun to Venus is 108,000,000 km. Calculate the time it takes for light to travel that distance.

Answer to Reading Check
The speed of light is about 880,000 times faster than the speed of sound.

Answer to Math Focus
1. *time* = 108,000,000 km ÷ 300,000 km/s = 360 s (= 6 min)

Light from the Sun

Even though light travels quickly, it takes about 8.3 min for light to travel from the sun to Earth. It takes this much time because Earth is 150,000,000 km away from the sun.

The EM waves from the sun are the major source of energy on Earth. For example, plants use photosynthesis to store energy from the sun. And animals use and store energy by eating plants or by eating other animals that eat plants. Even fossil fuels, such as coal and oil, store energy from the sun. Fossil fuels are formed from the remains of plants and animals that lived millions of years ago.

Although Earth receives a large amount of energy from the sun, only a very small part of the total energy given off by the sun reaches Earth. Look at **Figure 4.** The sun gives off energy as EM waves in all directions. Most of this energy travels away in space.

Figure 4 *Only a small amount of the sun's energy reaches the planets in the solar system.*

SECTION Review

Summary

- Light is an electromagnetic (EM) wave. An EM wave is a wave that consists of changing electric and magnetic fields. EM waves require no matter through which to travel.
- EM waves can be produced by the vibration of charged particles.
- The speed of light in a vacuum is about 300,000,000 m/s.
- EM waves from the sun are the major source of energy for Earth.

Using Key Terms

1. Use the following terms in the same sentence: *electromagnetic wave* and *radiation*.

Understanding Key Ideas

2. Electromagnetic waves are different from other types of waves because they can travel through
   - **a.** air.
   - **c.** space.
   - **b.** glass.
   - **d.** steel.

3. Describe light in terms of electromagnetic waves.

4. Why is light from the sun important?

5. How can electromagnetic waves be produced?

Math Skills

6. The distance from the sun to Jupiter is 778,000,000 km. How long does it take for light from the sun to reach Jupiter?

Critical Thinking

7. **Making Inferences** Why is it important that EM waves can travel through empty space?

8. **Making Comparisons** How does the amount of energy produced by the sun compare with the amount of energy that reaches Earth from the sun?

9. **Applying Concepts** Explain why the energy produced by burning wood in a campfire is energy from the sun.

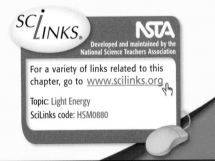

SciLINKS NSTA

Developed and maintained by the National Science Teachers Association

For a variety of links related to this chapter, go to www.scilinks.org

Topic: Light Energy
SciLinks code: HSM0880

Homework — GENERAL

Writing **Seeing the Past** Physicist Paul Hewitt has stated that all that we see, even our own reflection, is from the past. Have students write about whether they think that this statement is true, and why or why not. (If students have difficulty understanding Hewitt's statement, give them the following example: The light from stars in the sky may take years to reach Earth. In fact, it is possible that some of the stars that we see now no longer exist!) **LS Verbal**

CHAPTER RESOURCES

Chapter Resource File

- Section Quiz GENERAL
- Section Review GENERAL
- Vocabulary and Section Summary GENERAL

Focus

Overview

In this section, students will learn how electromagnetic waves differ from each other, and the role that different kinds of electromagnetic waves play in everyday life.

 Bellringer

Ask students to describe the weather conditions necessary to see a rainbow. (It has to be raining, and the sun must be shining.) Then, ask why rainbows form.

Motivate

Discussion ——— GENERAL

Wavelengths Discuss with students the wavelengths of various EM waves. Explain that radio waves broadcast by a typical FM radio station are about 3 m long. Have students use metersticks to measure and draw an FM radio wave of 3 m on the board. Then, explain that the waves produced by the typical microwave oven have a wavelength of about 12 cm. Have students draw a wave with a 12 cm wavelength. Challenge students to imagine how small the wavelength of visible light is, which ranges between 0.0000004 and 0.0000007 m! **LS** Visual

---

READING WARM-UP

Objectives
- Identify how electromagnetic waves differ from each other.
- Describe some uses for radio waves and microwaves.
- List examples of how infrared waves and visible light are important in your life.
- Explain how ultraviolet light, X rays, and gamma rays can be both helpful and harmful.

Terms to Learn
electromagnetic spectrum

READING STRATEGY

Mnemonics As you read this section, create a mnemonic device to help you remember the kinds of EM waves.

The Electromagnetic Spectrum

When you look around, you can see things that reflect light to your eyes. But a bee might see the same things differently. Bees can see a kind of light—called ultraviolet light—*that you can't see!*

It might seem odd to call something you can't see *light*. The light you are most familiar with is called *visible light*. Ultraviolet light is similar to visible light. Both are kinds of electromagnetic (EM) waves. In this section, you will learn about many kinds of EM waves, including X rays, radio waves, and microwaves.

Characteristics of EM Waves

All EM waves travel at the same speed in a vacuum—300,000 km/s. How is this possible? The speed of a wave is found by multiplying its wavelength by its frequency. So, EM waves having different wavelengths can travel at the same speed as long as their frequencies are also different. The entire range of EM waves is called the **electromagnetic spectrum.** The electromagnetic spectrum is shown in **Figure 1.** The electromagnetic spectrum is divided into regions according to the length of the waves. There is no sharp division between one kind of wave and the next. Some kinds even have overlapping ranges.

✔ **Reading Check** How is the speed of a wave determined? (*See the Appendix for answers to Reading Checks.*)

Figure 1 **The Electromagnetic Spectrum**

The electromagnetic spectrum is arranged from long to short wavelength or from low to high frequency.

Radio waves	**Microwaves**	**Infrared**
All radio and television stations broadcast radio waves.	Despite their name, microwaves are not the shortest EM waves.	*Infrared* means "below red."

CHAPTER RESOURCES

Chapter Resource File

- **Lesson Plan**
- **Directed Reading A** BASIC
- **Directed Reading B** SPECIAL NEEDS

Technology

Transparencies
- Bellringer
- The Electromagnetic Spectrum

Answer to Reading Check

The speed of a wave is determined by multiplying the wavelength and frequency of the wave.

Radio Waves

Radio waves cover a wide range of waves in the EM spectrum. Radio waves have some of the longest wavelengths and the lowest frequencies of all EM waves. In fact, radio waves are any EM waves that have wavelengths longer than 30 cm. Radio waves are used for broadcasting radio signals.

Broadcasting Radio Signals

Figure 2 shows how radio signals are broadcast. Radio stations encode sound information into radio waves by varying either the waves' amplitude or their frequency. Changing amplitude or frequency is called *modulation* (MAHJ uh LAY shuhn). You probably know that there are AM radio stations and FM radio stations. The abbreviation *AM* stands for "amplitude modulation," and the abbreviation *FM* stands for "frequency modulation."

Comparing AM and FM Radio Waves

AM radio waves are different from FM radio waves. For example, AM radio waves have longer wavelengths than FM radio waves do. And AM radio waves can bounce off the atmosphere and thus can travel farther than FM radio waves. But FM radio waves are less affected by electrical noise than AM radio waves are. So, music broadcast from FM stations sounds better than music broadcast from AM stations.

① A radio station converts sound into an electric current. The current produces radio waves that are sent out in all directions by the antenna.

② A radio receives radio waves and then converts them into an electric current, which is then converted to sound.

Figure 2 *Radio waves cannot be heard, but they can carry energy that can be converted into sound.*

electromagnetic spectrum all of the frequencies or wavelengths of electromagnetic radiation

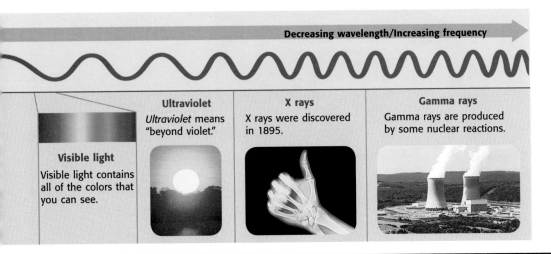

Decreasing wavelength/Increasing frequency

Visible light
Visible light contains all of the colors that you can see.

Ultraviolet
Ultraviolet means "beyond violet."

X rays
X rays were discovered in 1895.

Gamma rays
Gamma rays are produced by some nuclear reactions.

Is That a Fact!

Photoelectric cells, often found in alarm systems, change the light energy of a given frequency into an electric voltage. The electric current stops when the beam of light is broken, and this closes a default circuit. When the default circuit is triggered, it sets off an alarm.

Homework — BASIC

Poster Have students use the diagram of the EM spectrum in **Figure 1** to make a poster of the different parts of the spectrum. Students should include one or more examples of how we use each part of the spectrum. Encourage them to be creative and to illustrate their poster with drawings, photos, or images from magazines or newspapers. **LS** Visual

English Language Learners

Teach

READING STRATEGY — GENERAL

Prediction Guide Before students read this page, ask them, "If you were going to buy a radio station that would play mostly music, would you apply for an AM station license or an FM station license? Why? What if your station broadcast mostly news and sports? Which license would you apply for, and why?" Have students write their answers, and have them evaluate their answers after reading this page. **LS** Verbal

CONNECTION ACTIVITY
History — GENERAL

Writing **Light Scientists** Have students find out about the lives and accomplishments of the following pioneers in the study of EM waves:

• James Clerk Maxwell
• Heinrich Hertz
• Guglielmo Marconi
• Nikolai Tesla

Ask students to write a paragraph about what they learned.

(Maxwell introduced the theory of electromagnetic waves in 1865; Hertz created electromagnetic waves with a spark generator and reported that they could be transmitted over a distance; Marconi made the first wireless transmission across the Atlantic Ocean; Tesla developed the alternating current electrical system and some of the first electric motors.) **LS** Verbal

Research — GENERAL

Television Pioneers No one person is credited with the invention of television. Some pioneers of television technology include the German scientist Paul Nipkow (1860–1940), the American scientist and inventor Charles F. Jenkins (1867–1934), the Scottish inventor John Logie Baird (1888–1946), the Russian-born Vladymir Zworykin (1889–1982), the Japanese engineer Kenjiro Takayanagi (1899–1990), and American Philo T. Farnsworth (1906–1971). Have students research one of these developers of television technology and present what they learned in a oral report. **LS** Verbal

Answer to Reading Check

Radio waves carry TV signals.

CONNECTION ACTIVITY
Real Life — GENERAL

The Radio Spectrum The frequencies at which radio and television stations broadcast in the United States are assigned by the Federal Communications Commission (FCC). In fact, the FCC has assigned frequencies for all devices that use radio waves, including garage door openers, radio controlled toys, and baby monitors. Have students research how the FCC has divided the radio-wave spectrum and make a poster showing what they learned. **LS** Verbal

Radio Waves and Television

Television signals are also carried by radio waves. Most television stations broadcast radio waves that have shorter wavelengths and higher frequencies than those broadcast by radio stations. Like radio signals, television signals are broadcast using amplitude modulation and frequency modulation. Television stations use frequency-modulated waves to carry sound and amplitude-modulated waves to carry pictures.

Some waves carrying television signals are transmitted to artificial satellites orbiting Earth. The waves are amplified and sent to ground antennas. They then travel through cables to televisions in homes. Cable television works by this process.

✓ **Reading Check** Which EM waves can carry television signals?

Microwaves

Microwaves have shorter wavelengths and higher frequencies than radio waves do. Microwaves have wavelengths between 1 mm and 30 cm. You are probably familiar with microwaves—they are created in a microwave oven, such as the one shown in **Figure 3.**

Microwaves and Communication

Like radio waves, microwaves are used to send information over long distances. For example, cellular phones send and receive signals using microwaves. And signals sent between Earth and artificial satellites in space are also carried by microwaves.

Figure 3 How a Microwave Oven Works

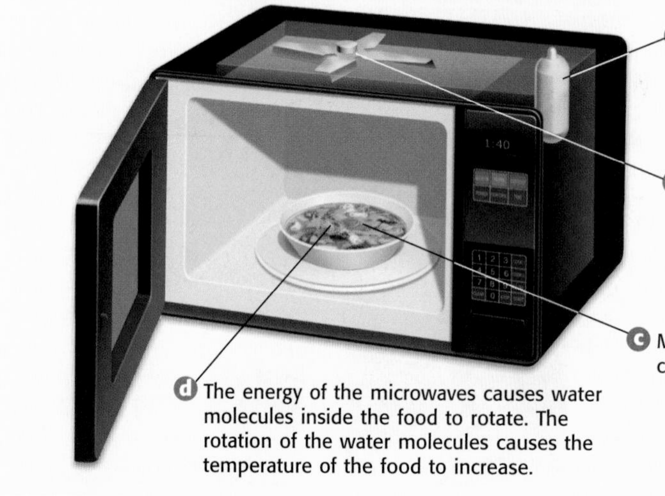

a A device called a *magnetron* produces microwaves by accelerating charged particles.

b The microwaves reflect off a metal fan and are directed into the cooking chamber.

c Microwaves can penetrate several centimeters into the food.

d The energy of the microwaves causes water molecules inside the food to rotate. The rotation of the water molecules causes the temperature of the food to increase.

Is That a Fact!

One of the first radio broadcasts in the United States occurred in 1910—a live concert of an opera featuring the great tenor Enrico Caruso.

WEIRD SCIENCE

When food is cooked in a microwave oven, the dish holding the food may get warm. But the microwave is not heating the dish. The dish is warmed by heat dissipating from the cooking food.

Figure 4 *Police officers use radar to detect cars going faster than the speed limit.*

Radar

Microwaves are also used in radar. *Radar* (**ra**dio **d**etection **a**nd **r**anging) is used to detect the speed and location of objects. The police officer in **Figure 4** is using radar to check the speed of a car. The radar gun sends out microwaves that reflect off the car and return to the gun. The reflected waves are used to calculate the speed of the car. Radar is also used to watch the movement of airplanes and to help ships navigate at night.

Infrared Waves

Infrared waves have shorter wavelengths and higher frequencies than microwaves do. The wavelengths of infrared waves vary between 700 nanometers and 1 mm. A nanometer (nm) is equal to 0.000000001 m.

On a sunny day, you may be warmed by infrared waves from the sun. Your skin absorbs infrared waves striking your body. The energy of the waves causes the particles in your skin to vibrate more, and you feel an increase in temperature. The sun is not the only source of infrared waves. Almost all things give off infrared waves, including buildings, trees, and you! The amount of infrared waves an object gives off depends on the object's temperature. Warmer objects give off more infrared waves than cooler objects do.

You can't see infrared waves, but some devices can detect infrared waves. For example, infrared binoculars change infrared waves into light you can see. Such binoculars can be used to watch animals at night. **Figure 5** shows a photo taken with film that is sensitive to infrared waves.

Figure 5 *In this photograph, brighter colors indicate higher temperatures.*

CONNECTION to
Earth Science ——— GENERAL

Infrared Film Scientists can detect mineral deposits, underground fires, diseased vegetation, and a variety of other things using infrared film. For instance, rocks and minerals vary in color because they are at slightly different temperatures. They appear a different color from each other on infrared film. In this way, geologists are able to chart the mineral content of the soil from the air. This technique lets scientists analyze soil on the moon and planets from spacecraft.

CONNECTION ACTIVITY
Life Science ——— GENERAL

Pit Vipers and Infrared Light
More than 140 species of snakes belong to the family known as pit vipers. The rattlesnake, cottonmouth, and copperhead are common North American pit vipers, and the bushmaster and fer-de-lance are pit vipers found in Central and South America. Pit vipers are unique in that they have a pair of organs called pits between the eyes and nostrils. The pits are very sensitive to infrared radiation up to 20 cm away. Nerve impulses from the pits and eyes are interpreted by the brain as a single picture. The pits allow pit vipers to see prey even at night. Have students research one pit viper and present what they learned in a poster or an oral report.
LS Verbal/Visual

WEIRD SCIENCE

Scientists have conducted experiments that slowed down a beam of laser light to 17 m/s (38 mph)! This incredible phenomenon was achieved by firing a laser beam through a gas cloud of sodium atoms cooled to only a fraction of a degree above absolute zero.

SCIENCE HUMOR

Q: How many actors does it take to change a light bulb?

A: Only one. They don't like to share the spotlight!

Rainbow Research Ask students, "What is a rainbow?" Basically, a rainbow is sunlight spread out into its spectrum of colors. The colors are directed toward the viewer by raindrops or other water droplets in the air. But rainbows are more complex than this. Challenge students to research how rainbows are formed and why we see them as we do. Students should present what they learned in an oral report or a paper. Encourage students to be creative; presentations might include models of rainbows, photographs of rainbows, or creating a rainbow in the school parking lot. **LS** Verbal

---

MISCONCEPTION /// ALERT \\\

The Visible Spectrum The range of the electromagnetic spectrum that humans can see is called the visible spectrum. However, other animals are able to see electromagnetic radiation with wavelengths outside of the visible spectrum. For example, bees and other insects use ultraviolet light, and pit vipers can detect infrared radiation.

Answer to Reading Check

White light is the combination of visible light of all wavelengths.

Figure 6 *Water droplets can separate white light into visible light of different wavelengths. As a result, you see all the colors of visible light in a rainbow.*

Making a Rainbow

On a sunny day, ask a parent to use a hose or a spray bottle to make a mist of water outside. Move around until you see a rainbow in the water mist. Draw a diagram showing the positions of the water mist, the sun, the rainbow, and yourself.

Visible Light

Visible light is the very narrow range of wavelengths and frequencies in the electromagnetic spectrum that humans can see. Visible light waves have shorter wavelengths and higher frequencies than infrared waves do. Visible light waves have wavelengths between 400 nm and 700 nm.

Visible Light from the Sun

Some of the energy that reaches Earth from the sun is visible light. The visible light from the sun is white light. *White light* is visible light of all wavelengths combined. Light from lamps in your home as well as from the fluorescent bulbs in your school is also white light.

✓ Reading Check What is white light?

Colors of Light

Humans see the different wavelengths of visible light as different colors, as shown in **Figure 6.** The longest wavelengths are seen as red light. The shortest wavelengths are seen as violet light.

The range of colors is called the *visible spectrum.* You can see the visible spectrum in **Figure 7.** When you list the colors, you might use the imaginary name *ROY G. BiV* to help you remember their order. The capital letters in Roy's name represent the first letter of each color of visible light: **r**ed, **o**range, **y**ellow, **g**reen, **b**lue, and **v**iolet. What about the *i* in Roy's last name? You can think of *i* as standing for the color indigo. Indigo is a dark blue color.

Is That a Fact!

John Dalton (1766–1844), an English chemist and researcher, had colorblindness. Dalton was one of the first scientist to study colorblindness. In 1794, Dalton presented a paper on colorblindness before the Manchester Literary and Philosophical Society. Dalton's paper was the first known description of this vision phenomenon, and for many years colorblindness was called Daltonism.

SCIENCE HUMOR

Q: Why did the beam of light look sad after meeting with the prism?

A: It was all broken up inside!

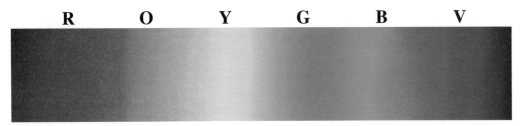

R O Y G B V

Figure 7 *The visible spectrum contains all colors of light.*

Ultraviolet Light

Ultraviolet light (UV light) is another type of electromagnetic wave produced by the sun. Ultraviolet waves have shorter wavelengths and higher frequencies than visible light does. The wavelengths of ultraviolet light waves vary between 60 nm and 400 nm. Ultraviolet light affects your body in both bad and good ways.

✓ **Reading Check** How do ultraviolet light waves compare with visible light waves?

Bad Effects

On the bad side, too much ultraviolet light can cause sunburn, as you can see in **Figure 8.** Too much ultraviolet light can also cause skin cancer, wrinkles, and damage to the eyes. Luckily, much of the ultraviolet light from the sun does not reach Earth's surface. But you should still protect yourself against the ultraviolet light that does reach you. To do so, you should use sunscreen with a high SPF (**s**un **p**rotection **f**actor). You should also wear sunglasses that block out UV light to protect your eyes. Hats, long-sleeved shirts, and long pants can protect you, too. You need this protection even on overcast days because UV light can travel through clouds.

Figure 8 *Too much exposure to ultraviolet light can lead to a painful sunburn. Using sunscreen will help protect your skin.*

Good Effects

On the good side, ultraviolet waves produced by ultraviolet lamps are used to kill bacteria on food and surgical tools. In addition, small amounts of ultraviolet light are beneficial to your body. When exposed to ultraviolet light, skin cells produce vitamin D. This vitamin allows the intestines to absorb calcium. Without calcium, your teeth and bones would be very weak.

CONNECTION to Real World — GENERAL

Phototherapy Therapy involving exposure to certain kinds of light has become an accepted treatment of a mood disorder known as *seasonal affective disorder.* Seasonal affective disorder, or SAD, is characterized by feelings of depression that typically occur during the fall and winter months. Researchers think that the reduction in the amount of light that passes through the eyes during these months affects the release of important brain chemicals. The treatment, known as *phototherapy,* involves a 20- to 30- minute daily exposure to a specific kind of light.

Answer to Reading Check
Ultraviolet light waves have shorter wavelengths and higher frequencies than visible light waves do.

CONNECTION ACTIVITY Math — GENERAL

SPF Numbers Sunscreens are ranked by sun protection factor (SPF) numbers. The numbers indicate how many times longer than normal that a person can stay in the sun without getting a sunburn. For example, suppose a person normally burns after being in the sun for 10 min. If he or she used sunscreen with SPF 15, he or she could stay in the sun for 150 min without getting burned. Have students calculate how long a person wearing sunscreen with SPF 15 can stay in the sun if he or she normally burns after being in the sun for 15 min. (225 min) Then, ask students why people who burn easily need a higher SPF. (People who burn easily also tend to burn quickly. Therefore, they need a higher SPF to be able to stay in the sun for a long time without getting burned.) **LS** Logical

ACTIVITY — BASIC

Blocking UV Light Organize students into groups. Give each group a few coins and a piece of construction paper. Each piece of paper should be a different color. Have each group place the coins on their paper and set the paper outside in direct sunlight. (Glass windows will block UV light.) Have students wait a couple of days and compare the amounts of fading among the different colored papers. Explain to students that the UV light breaks down different types of dye at different rates. **LS** Kinesthetic

CONNECTION to Life Science — GENERAL

Vitamin D Our bodies can make vitamin D if the skin is exposed to ultraviolet light in sunlight for certain periods of time. Many people are not exposed to sunlight long enough for their bodies to produce enough vitamin D. However, vitamin D is available in some foods. Good dietary sources of vitamin D are milk and dairy products, butter, eggs, liver, cod-liver oil, and oily fish, such as salmon.

Reteaching — BASIC

Wavelength and Frequency To clarify the relationship between wavelength and frequency, draw a wave across the board with a gradually decreasing wavelength. Then, slowly move the pointer at a constant speed across the wave, and ask students to clap each time the pointer touches a crest of the wave. Discuss with students how the frequency of their claps increased as the wavelength decreased. **LS Kinesthetic**

Quiz — GENERAL

1. What is the electromagnetic spectrum? (The electromagnetic spectrum is the entire range of electromagnetic waves.)

2. List four different kinds of electromagnetic waves. (Answers should include four of the following: radio waves, microwaves, infrared light, visible light, ultraviolet light, X rays, and gamma rays.)

3. True or false: FM radio waves can travel greater distances than AM radio waves. (false)

Alternative Assessment — GENERAL

Writing **Short Story** Have students write a short story in which the characters use or are affected by each of the different kinds of EM waves. **LS Verbal**

CONNECTION TO Astronomy

Gamma Ray Spectrometer In 2001, NASA put an artificial satellite called the *2001 Mars Odyssey* in orbit around Mars. The *Odyssey* is carrying a gamma ray spectrometer. A *spectrometer* is a device used to detect certain kinds of EM waves. The gamma ray spectrometer on the *Odyssey* was used to look for water and several chemical elements on Mars. Scientists hope to use this information to learn about the geology of Mars. Research the characteristics of Mars and Earth. In your **science journal,** make a chart comparing Mars and Earth.

ACTIVITY

X Rays and Gamma Rays

X rays and gamma rays have some of the shortest wavelengths and highest frequencies of all EM waves.

X Rays

X rays have wavelengths between 0.001 nm and 60 nm. They can pass through many materials. This characteristic makes X rays useful in the medical field, as shown in **Figure 9.** But too much exposure to X rays can also damage or kill living cells. A patient getting an X ray may wear special aprons to protect parts of the body that do not need X-ray exposure. These aprons are lined with lead because X rays cannot pass through lead.

X-ray machines are also used as security devices in airports and other public buildings. The machines allow security officers to see inside bags and other containers without opening the containers.

✓ Reading Check How are patients protected from X rays?

Gamma Rays

Gamma rays are EM waves that have wavelengths shorter than 0.1 nm. They can penetrate most materials very easily. Gamma rays are used to treat some forms of cancer. Doctors focus the rays on tumors inside the body to kill the cancer cells. This treatment often has good effects, but it can have bad side effects because some healthy cells may also be killed.

Gamma rays are also used to kill harmful bacteria in foods, such as meat and fresh fruits. The gamma rays do not harm the treated food and do not stay in the food. So, food that has been treated with gamma rays is safe for you to eat.

Figure 9 How a Bone Is X Rayed

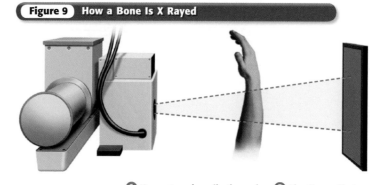

❶ X rays travel easily through skin and muscle but are absorbed by bones.

❷ The X rays that are not absorbed strike the film.

❸ Bright areas appear on the film where X rays are absorbed by the bones.

Is That a Fact!

Cosmic photons are a kind of EM wave. These photons have greater frequencies and shorter wavelengths than gamma rays. The incredible energy required to create these cosmic photons may come from supernovas or other astrophysical phenomena. Refer to the teaching transparencies "The H-R Diagram: A and B" to help students understand supernovas.

Answer to Reading Check

Patients are protected from X rays by special lead-lined aprons.

Summary

- All electromagnetic (EM) waves travel at the speed of light. EM waves differ only by wavelength and frequency.
- The entire range of EM waves is called the *electromagnetic spectrum*.
- Radio waves are used for communication.
- Microwaves are used in cooking and in radar.
- The absorption of infrared waves is felt as an increase in temperature.
- Visible light is the narrow range of wavelengths that humans can see. Different wavelengths are seen as different colors.
- Ultraviolet light is useful for killing bacteria and for producing vitamin D in the body. Overexposure to ultraviolet light can cause health problems.
- X rays and gamma rays are EM waves that are often used in medicine. Overexposure to these kinds of rays can damage or kill living cells.

Using Key Terms

1. In your own words, write a definition for the term *electromagnetic spectrum*.

Understanding Key Ideas

2. Which of the following electromagnetic waves are produced by the sun?
   a. infrared waves
   b. visible light
   c. ultraviolet light
   d. All of the above

3. How do the different kinds of EM waves differ from each other?

4. Describe two ways of transmitting information using radio waves.

5. Explain why ultraviolet light, X rays, and gamma rays can be both helpful and harmful.

6. What are two common uses for microwaves?

7. What is white light? What are two sources of white light?

8. What is the visible spectrum?

Critical Thinking

9. **Applying Concepts** Describe how three different kinds of electromagnetic waves have been useful to you today.

10. **Making Comparisons** Compare the wavelengths of infrared waves, ultraviolet light, and visible light.

Interpreting Graphics

The waves in the diagram below represent two different kinds of EM waves. Use the diagram below to answer the questions that follow.

11. Which wave has the longest wavelength?

12. Suppose that one of the waves represents a microwave and one of the waves represents a radio wave. Which wave represents the microwave?

CHAPTER RESOURCES

Chapter Resource File

- Section Quiz **GENERAL**
- Section Review **GENERAL**
- Vocabulary and Section Summary **GENERAL**
- SciLinks Activity **GENERAL**

Technology

Transparencies
- **LINK TO EARTH SCIENCE** The H-R Diagram: A and B

Answers to Section Review

1. Sample answer: The electromagnetic spectrum is made up of electromagnetic waves of all wavelengths.

2. d

3. The different kinds of EM waves differ by their wavelengths and frequencies.

4. Radio waves can transmit information by varying the waves' amplitude (called amplitude modulation) or by varying the waves' frequency (called frequency modulation).

5. UV light is useful because it can kill bacteria and help the human body produce vitamin D. But overexposure to UV light can cause sunburn and skin cancer. X rays and gamma rays are used in medicine to check for broken bones and to treat cancer. But X rays and gamma rays can also kill healthy, living cells.

6. Sample answer: Two common uses for microwaves are cooking in a microwave oven and checking the speeds of cars using radar.

7. White light is visible light of all wavelengths combined. Sources of white light include the sun, incandescent light bulbs, and fluorescent light bulbs.

8. The entire range of colors of visible light is the visible spectrum.

9. Accept all reasonable answers. Sample answer: Visible light helped me see so I didn't walk into things, radio waves transmitted television signals so I could watch TV, and microwaves warmed my food.

10. Infrared light has the longest wavelength, visible light has the next longest, and ultraviolet light has the shortest wavelength.

11. Wave **a** has the longest wavelength.

12. Wave **b** represents the microwave, because microwaves have shorter wavelengths than radio waves.

Focus

Overview

This section discusses reflection, refraction, diffraction, and inter-ference of light waves. Students learn how light is absorbed and scattered and how white light can be separated into colors.

 Bellringer

Tell students that mirrors are common objects that most people use every day. Then ask students, "From your experience, how do mirrors work and what do mirrors do to light waves?"

Motivate

Group ACTIVITY—GENERAL

Making a Periscope Organize students into groups. Give each group a shoe box, two small hand mirrors, some modeling clay, and a pair of scissors. Have students cut a 3 cm hole on the left side of each end of the box (so the holes will not be directly opposite each other). Then, tell students to arrange the mirrors inside of the box with the modeling clay in such a way that someone can look straight into one hole and see out of the other hole. Ask students to explain how this device works. **LS** Kinesthetic

---

READING WARM-UP

Objectives

- Describe how reflection allows you to see things.
- Describe absorption and scattering.
- Explain how refraction can create optical illusions and separate white light into colors.
- Explain the relationship between diffraction and wavelength.
- Compare constructive and destructive interference of light.

Terms to Learn

reflection   refraction
absorption   diffraction
scattering   interference

READING STRATEGY

Reading Organizer As you read this section, make a concept map by using the terms above.

Interactions of Light Waves

Have you ever seen a cat's eyes glow in the dark when light shines on them? Cats have a special layer of cells in the back of their eyes that reflects light.

This layer helps the cat see better by giving the eyes another chance to detect the light. Reflection is one interaction of electromagnetic waves. Because we can see visible light, it is easier to explain all wave interactions by using visible light.

Reflection

Reflection happens when light waves bounce off an object. Light reflects off objects all around you. When you look in a mirror, you are seeing light that has been reflected twice—first from you and then from the mirror. If light is reflecting off everything around you, why can't you see your image on a wall? To answer this question, you must learn the law of reflection.

The Law of Reflection

Light reflects off surfaces the same way that a ball bounces off the ground. If you throw the ball straight down against a smooth surface, it will bounce straight up. If you bounce it at an angle, it will bounce away at an angle. The *law of reflection* states that the angle of incidence is equal to the angle of reflection. *Incidence* is the arrival of a beam of light at a surface. **Figure 1** shows this law.

 **Reading Check** What is the law of reflection? (*See the Appendix for answers to Reading Checks.*)

Figure 1 The Law of Reflection

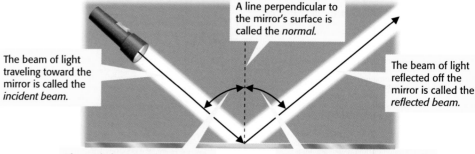

A line perpendicular to the mirror's surface is called the *normal*.

The beam of light traveling toward the mirror is called the *incident beam*.

The beam of light reflected off the mirror is called the *reflected beam*.

The angle between the incident beam and the normal is called the *angle of incidence*.

The angle between the reflected beam and the normal is called the *angle of reflection*.

CHAPTER RESOURCES

Chapter Resource File

 • Lesson Plan
- Directed Reading A **BASIC**
- Directed Reading B **SPECIAL NEEDS**

Technology

 Transparencies
- Bellringer
- The Law of Reflection
- Regular Reflection Versus Diffuse Reflection

Answer to Reading Check

The law of reflection states that the angle of incidence equals the angle of reflection.

Figure 2 Regular Reflection Vs. Diffuse Reflection

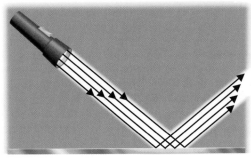

Regular reflection occurs when light beams are reflected at the same angle. When your eye detects the reflected beams, you can see a reflection on the surface.

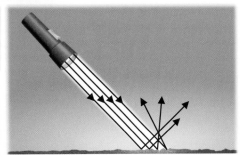

Diffuse reflection occurs when light beams reflect at many different angles. You can't see a reflection because not all of the reflected light is directed toward your eyes.

Types of Reflection

So, why can you see your image in a mirror but not in a wall? The answer has to do with the differences between the two surfaces. A mirror's surface is very smooth. Thus, light beams reflect off all points of the mirror at the same angle. This kind of reflection is called *regular reflection*. A wall's surface is slightly rough. Light beams will hit the wall's surface and reflect at many different angles. This kind of reflection is called *diffuse reflection*. **Figure 2** shows the difference between the two kinds of reflection.

Light Source or Reflection?

If you look at a TV set in a bright room, you see the cabinet around the TV and the image on the screen. But if you look at the same TV in the dark, you see only the image on the screen. The difference is that the screen is a light source, but the cabinet around the TV is not.

You can see a light source even in the dark because its light passes directly into your eyes. The tail of the firefly in **Figure 3** is a light source. Flames, light bulbs, and the sun are also light sources. Objects that produce visible light are called *luminous* (LOO muh nuhs).

Most things around you are not light sources. But you can still see them because light from light sources reflects off the objects and then travels to your eyes. A visible object that is not a light source is *illuminated*.

 **Reading Check** List four different light sources.

reflection the bouncing back of a ray of light, sound, or heat when the ray hits a surface that it does not go through

Figure 3 *You can see the tail of this firefly because it is luminous. But you see its body because it is illuminated.*

ACTIVITY ———— ADVANCED

Why Is the Sky Blue? Have students research why the sky appears blue. Wrong answers to this question are quite common. Students should find that when visible light from the sun passes through the atmosphere, air molecules scatter light waves with shorter wavelengths. Blue happens to be the most affected wavelength. Encourage students to create posters or demonstrations to explain what they learned. **LS** Verbal

CONNECTION TO Astronomy

Moonlight? Sometimes, the moon shines so brightly that you might think there is a lot of "moonlight." But did you know that moonlight is actually sunlight? The moon does not give off light. You can see the moon because it is illuminated by light from the sun. You see different phases of the moon because light from the sun shines only on the part of the moon that faces the sun. Make a poster that shows the different phases of the moon. **ACTIVITY**

absorption in optics, the transfer of light energy to particles of matter

scattering an interaction of light with matter that causes light to change its energy, direction of motion, or both

Absorption and Scattering

Have you noticed that when you use a flashlight, the light shining on things closer to you appears brighter than the light shining on things farther away? The light is less bright the farther it travels from the flashlight. The light is weaker partly because the beam spreads out and partly because of absorption and scattering.

Absorption of Light

The transfer of energy carried by light waves to particles of matter is called **absorption.** When a beam of light shines through the air, particles in the air absorb some of the energy from the light. As a result, the beam of light becomes dim. The farther the light travels from its source, the more it is absorbed by particles, and the dimmer it becomes.

Scattering of Light

Scattering is an interaction of light with matter that causes light to change direction. Light scatters in all directions after colliding with particles of matter. Light from the ship shown in **Figure 4** is scattered out of the beam by air particles. This scattered light allows you to see things that are outside the beam. But, because light is scattered out of the beam, the beam becomes dimmer.

Scattering makes the sky blue. Light with shorter wavelengths is scattered more than light with longer wavelengths. Sunlight is made up of many different colors of light, but blue light (which has a very short wavelength) is scattered more than any other color. So, when you look at the sky, you see a background of blue light.

✓ **Reading Check** Why can you see things outside a beam of light?

Figure 4 *A beam of light becomes dimmer partly because of scattering.*

Is That a Fact!

A blue jay's feathers are not really blue. The air molecules on the surface barbs of the feathers scatter the red and green light of the visible spectrum, leaving blue light to reflect to our eyes.

Answer to Reading Check

You can see things outside of a beam of light because light is scattered outside of the beam.

Scattering Milk

1. Fill a **2 L clear plastic bottle** with **water**.
2. Turn the lights off, and shine a **flashlight** through the water. Look at the water from all sides of the bottle. Write a description of what you see.
3. Add **3 drops of milk** to the water, and shake the bottle to mix it up.
4. Repeat step 2. Describe any color changes. If you don't see any, add more milk until you do.
5. How is the water-and-milk mixture like air particles in the atmosphere? Explain your answer.

Refraction

Imagine that you and a friend are at a lake. Your friend wades into the water. You look at her, and her feet appear to have separated from her legs! What has happened? You know her feet did not fall off, so how can you explain what you see? The answer has to do with refraction.

Refraction and Material

Refraction is the bending of a wave as it passes at an angle from one substance, or material, to another. **Figure 5** shows a beam of light refracting twice. Refraction of light waves occurs because the speed of light varies depending on the material through which the waves are traveling. In a vacuum, light travels at 300,000 km/s, but it travels more slowly through matter. When a wave enters a new material at an angle, the part of the wave that enters first begins traveling at a different speed from that of the rest of the wave.

refraction the bending of a wave as the wave passes between two substances in which the speed of the wave differs

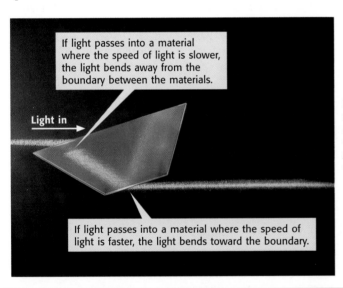

If light passes into a material where the speed of light is slower, the light bends away from the boundary between the materials.

Light in

If light passes into a material where the speed of light is faster, the light bends toward the boundary.

Figure 5 *Light travels more slowly through glass than it does through air. So, light refracts as it passes at an angle from air to glass or from glass to air. Notice that the light is also reflected inside the prism.*

Early Mirrors The earliest mirrors date back to 6000 BCE and were discovered in areas around Turkey and Egypt. These mirrors were about 90 mm in diameter and were made from flat pieces of polished igneous rock called obsidian.

Is That a Fact!

Some light bulbs and mirrors are frosted or etched to reduce glare or to create interesting visual effects. This process involves applying hydrofluoric acid, a powerful and toxic acid, directly to the glass. This acid reacts with silicon atoms in the glass to produce this effect.

Archerfish Archerfish *(Toxotes jaculator)* have the ability to correct for the refraction of light between air and water, and they have the ability to judge the distance to their prey. There are six species of archerfish that live in estuaries, wetlands, and fresh water in Southeast Asia, the western Pacific, and Australia. Archerfish can grow to 41 cm in length. Archerfish knock insects and other small prey from overhanging vegetation by spitting jets of water at them. Archerfish have a groove in the roof of their mouth. When the tongue is pressed against the groove and the gills are squeezed shut, a jet of water is produced. These fish can hit prey more than 1.5 m away! Challenge students to demonstrate how difficult it is to correct for the refraction of light between air and water. (Sample answer: Students may place an object underwater in a fish bowl and ask other students to try to touch the object with a pencil while looking through the surface of the water at an angle.) **LS** **Logical/Kinesthetic**

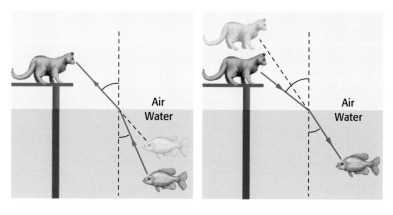

Figure 6 *Because of refraction, the cat and the fish see optical illusions. To the cat, the fish appears closer than it really is. To the fish, the cat appears farther away than it actually is.*

Refraction and Optical Illusions

Usually, when you look at an object, the light reflecting off the object travels in a straight line from the object to your eye. Your brain always interprets light as traveling in straight lines. But when you look at an object that is underwater, the light reflecting off the object does not travel in a straight line. Instead, it refracts. **Figure 6** shows how refraction creates an optical illusion. This kind of illusion causes a person's feet to appear separated from the legs when the person is wading.

Refraction and Color Separation

White light is composed of all the wavelengths of visible light. The different wavelengths of visible light are seen by humans as different colors. When white light is refracted, the amount that the light bends depends on its wavelength. Waves with short wavelengths bend more than waves with long wavelengths. As shown in **Figure 7,** white light can be separated into different colors during refraction. Color separation by refraction is responsible for the formation of rainbows. Rainbows are created when sunlight is refracted by water droplets.

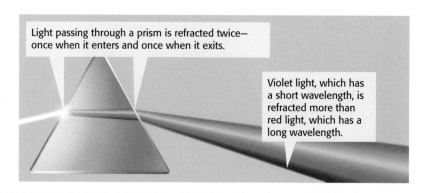

Light passing through a prism is refracted twice— once when it enters and once when it exits.

Figure 7 *A prism is a piece of glass that separates white light into the colors of visible light by refraction.*

Violet light, which has a short wavelength, is refracted more than red light, which has a long wavelength.

Rainbow Theory The first person to explain how rainbows form was René Descartes (1596–1650), a famous French mathematician and scientist. His explanation was published in 1637 in his book *Discours de la Methode.* Descartes reasoned that because rainbows also appear in waterfalls and water fountains, as well as in the sky, rainbows must be a result of drops of water affecting light waves.

Seeing Rainbows Natural rainbows are visible only when the sun is lower than 42° above the horizon. The reason is related to the way water refracts through raindrops. It is unlikely that you will see a rainbow at noon!

Refraction Rainbow

1. **Tape** a **piece of construction paper** over the end of a **flashlight.** Use **scissors** to cut a slit in the paper.

2. Turn on the flashlight, and lay it on a table. Place a **prism** on end in the beam of light.

3. Slowly rotate the prism until you can see a rainbow on the surface of the table. Draw a diagram of the light beam, the prism, and the rainbow.

Diffraction

Refraction isn't the only way light waves are bent. **Diffraction** is the bending of waves around barriers or through openings. The amount a wave diffracts depends on its wavelength and the size of the barrier or the opening. The greatest amount of diffraction occurs when the barrier or opening is the same size or smaller than the wavelength.

diffraction a change in the direction of a wave when the wave finds an obstacle or an edge, such as an opening

✔ **Reading Check** The amount a wave diffracts depends on what two things?

Diffraction and Wavelength

The wavelength of visible light is very small—about 100 times thinner than a human hair! So, a light wave cannot bend very much by diffraction unless it passes through a narrow opening, around sharp edges, or around a small barrier, as shown in **Figure 8.**

Light waves cannot diffract very much around large obstacles, such as buildings. Thus, you can't see around corners. But light waves always diffract a small amount. You can observe light waves diffracting if you examine the edges of a shadow. Diffraction causes the edges of shadows to be blurry.

Figure 8 *This diffraction pattern is made by light of a single wavelength shining around the edges of a very tiny disk.*

Answer to Reading Check

The amount that a wave diffracts depends on the wavelength of the wave and the size of the barrier or opening.

MATERIALS

FOR EACH GROUP
- flashlight
- paper, construction
- prism
- scissors
- tape

Teacher's Notes: For best results, this activity should be done in a darkened room with powerful flashlights. Alternatively, this activity can be done outside by using sunlight instead of a flashlight.

Answer

3. Accept all reasonable answers. Students' diagrams should show that light enters on one side of the prism and exits on a different side, producing a rainbow. Their drawings should also show that violet light refracts more than red light.

Demonstration —GENERAL

Diffraction Obtain a piece of diffraction grating and a laser. Darken the room, and point the laser at a wall. Ask the students to describe what they see on the wall. (a single red dot)

Place the diffraction grating in front of the laser light, and aim the beam at the same wall. Ask students what they see on the wall. (a series of small red dots)

Have students look at the dots carefully and describe what they see. (The dots aren't really round like the single dot.)

Explain to students that a diffraction grating is a piece of plastic or glass that has thin, closely spaced lines on it. Discuss diffraction with students and how diffracted laser light causes the series of flattened circles of light instead of a single dot. **LS Visual**

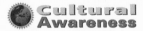

Cultural Awareness GENERAL

Speed of Light Experiment The French scientist Hippolyte-Louis Fizeau was the first person to measure the speed of light using a laboratory experiment. Fizeau used mirrors and a rotating, toothed wheel to break a beam of light into a series of pulses and then measure the speed of those pulses.

Close

Reteaching — BASIC

Reflecting Balls Give each pair of students a ball that bounces. Have students bounce the ball to each other at different angles. Explain to students that the way the balls bounce is similar to the law of reflection. **LS Kinesthetic**

Quiz — GENERAL

1. The moon does not produce light of its own. So, where does moonlight come from? (Moonlight is sunlight that is reflected off the moon.)

2. Archerfish shoot jets of water at insects sitting on vegetation above the water, so they must adjust for refraction. Where is the ideal place for an archerfish to be in relation to its prey? Why? (The archerfish should be directly below its prey. Light entering the water perpendicular to the surface is not refracted.)

3. Explain how colors are separated in a rainbow. (Rainbows form when water droplets refract sunlight. Light of shorter wavelengths, such as violet, is refracted more than light of longer wavelengths, such as red.)

Alternative Assessment — GENERAL

Concept Mapping Have students make a concept map about the interactions of light waves. **LS Verbal/Visual**

INTERNET ACTIVITY

For another activity related to this chapter, go to **go.hrw.com** and type in the keyword **HP5LGTW**.

---

interference the combination of two or more waves that results in a single wave

Interference

Interference is a wave interaction that happens when two or more waves overlap. Overlapping waves can combine by constructive or destructive interference.

Constructive Interference

When waves combine by *constructive interference,* the resulting wave has a greater amplitude, or height, than the individual waves had. Constructive interference of light waves can be seen when light of one wavelength shines through two small slits onto a screen. The light on the screen will appear as a series of alternating bright and dark bands, as shown in **Figure 9.** The bright bands result from light waves combining through constructive interference.

✓ **Reading Check** What is constructive interference?

Destructive Interference

When waves combine by *destructive interference,* the resulting wave has a smaller amplitude than the individual waves had. So, when light waves interfere destructively, the result will be dimmer light. Destructive interference forms the dark bands seen in **Figure 9.**

You do not see constructive or destructive interference of white light. To understand why, remember that white light is composed of waves with many different wavelengths. The waves rarely line up to combine in total destructive interference.

Figure 9 **Constructive and Destructive Interference**

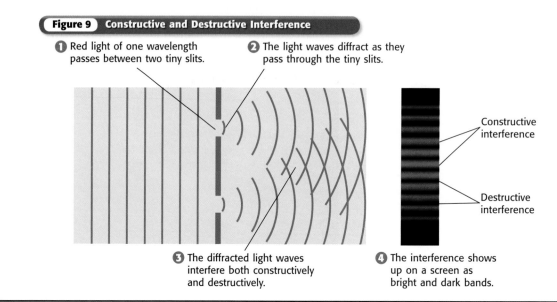

❶ Red light of one wavelength passes between two tiny slits.

❷ The light waves diffract as they pass through the tiny slits.

❸ The diffracted light waves interfere both constructively and destructively.

❹ The interference shows up on a screen as bright and dark bands.

Constructive interference

Destructive interference

CONNECTION ACTIVITY Real World — GENERAL

Holograms A common use of light interference is using laser light to create a hologram. Show students several examples of holograms. Ask students to compare holograms with photographs. (Sample answers: Photographs can be in color; holograms are not colored. Images in photographs are flat; images in holograms are three-dimensional.) **LS Visual**

Answer to Reading Check

Constructive interference is interference in which the resulting wave has a greater amplitude than the original waves had.

SECTION Review

Summary

- The law of reflection states that the angle of incidence is equal to the angle of reflection.
- Things that are luminous can be seen because they produce their own light. Things that are illuminated can be seen because light reflects off them.
- Absorption is the transfer of light energy to particles of matter. Scattering is an interaction of light with matter that causes light to change direction.

- Refraction of light waves can create optical illusions and can separate white light into separate colors.
- How much light waves diffract depends on the light's wavelength. Light waves diffract more when traveling through a narrow opening.
- Interference can be constructive or destructive. Interference of light waves can cause bright and dark bands.

Using Key Terms

For each pair of terms, explain how the meanings of the terms differ.

1. *refraction* and *diffraction*
2. *absorption* and *scattering*

Understanding Key Ideas

3. Which light interaction explains why you can see things that do not produce their own light?
   - **a.** absorption
   - **b.** reflection
   - **c.** refraction
   - **d.** scattering

4. Describe how absorption and scattering can affect a beam of light.
5. Why do objects that are underwater look closer than they actually are?
6. How does a prism separate white light into different colors?
7. What is the relationship between diffraction and the wavelength of light?

Critical Thinking

8. **Applying Concepts** Explain why you can see your reflection on a spoon but not on a piece of cloth.
9. **Making Inferences** The planet Mars does not produce light. Explain why you can see Mars shining like a star at night.

10. **Making Comparisons** Compare constructive interference and destructive interference.

Interpreting Graphics

Use the image below to answer the questions that follow.

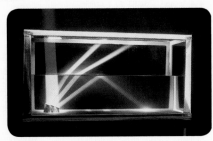

11. Why doesn't the large beam of light bend like the two beams in the middle of the tank?
12. Which light interaction explains what is happening to the bottom light beam?

SCI LINKS

NSTA
Developed and maintained by the National Science Teachers Association

For a variety of links related to this chapter, go to www.scilinks.org

Topic: Reflection and Refraction
SciLinks code: HSM1283

Answers to Section Review

1. Refraction is the bending of waves when the waves pass at an angle from one material to another. Diffraction is the bending of waves around barriers or through openings.

2. Absorption is the transfer of light energy to particles of matter. Scattering is an interaction of light with matter that causes light to change its energy, direction of motion, or both.

3. b

CHAPTER RESOURCES

Chapter Resource File

- Section Quiz **GENERAL**
- Section Review **GENERAL**
- Vocabulary and Section Summary **GENERAL**
- Reinforcement Worksheet **BASIC**
- Critical Thinking **ADVANCED**
- Datasheet for Quick Lab

---

4. Absorption and scattering can make a beam of light dimmer. As light travels through the air, particles in the air absorb some of the energy from the light. As a result, the beam of light becomes dim. Scattering causes light to change direction and move out of the beam of light. This also causes the beam to become dimmer.

5. Objects underwater look closer than they actually are because refraction creates an optical illusion.

6. A prism separates white light into different colors by refraction. Light waves that have shorter wavelengths refract more than light waves that have longer wavelengths. The difference in the amount of refraction separates the colors of light.

7. The greatest amount of diffraction occurs when the barrier around which the light is traveling (or the opening through which the light is traveling) is the same size or smaller than the wavelength of light.

8. Light reflects off a spoon by regular reflection, so you can see your image in the spoon. But light reflects off a piece of cloth by diffuse reflection. Thus, you can see the cloth, but not your image.

9. You can see Mars because light from the sun reflects off it.

10. When waves combine by constructive interference, the resulting wave has a larger amplitude than the individual waves had. When waves combine by destructive interference, the resulting wave has a smaller amplitude than the individual waves had.

11. The large beam of light does not bend because it is not at an angle. Refraction only occurs when light waves pass from one substance to another at an angle.

12. The bottom light beam is reflecting off the surface of the water.

Focus

Overview

Students will learn how light interacts with matter. Students will also learn what determines the color of an object, and they will learn about mixing colors of light and mixing pigments.

 **Bellringer**

Ask students the following: "What is your favorite color? In a short paragraph, explain why you like your favorite color. Also, explain how certain colors affect your mood."

Motivate

Demonstration — GENERAL

Adding Colors Cover one lens of a high-intensity flashlight with a green filter and a second flashlight lens with a red filter. In a darkened room, turn the "green" light on, and shine it on a white sheet, white wall, or overhead screen. Turn the "red" light on, and shine it on a different area of the screen. Ask the students what colors they see. Ask students to predict what color they will see if you overlap the green and red light. The students will probably answer "brown." Overlap the two colors. (Yellow will appear.) **LS Visual**

READING WARM-UP

Objectives

- Name and describe the three ways light interacts with matter.
- Explain how the color of an object is determined.
- Explain why mixing colors of light is called *color addition*.
- Describe why mixing colors of pigments is called *color subtraction*.

Terms to Learn

transmission   opaque
transparent   pigment
translucent

READING STRATEGY

Discussion Read this section silently. Write down questions that you have about this section. Discuss your questions in a small group.

transmission the passing of light or other form of energy through matter

---

Light and Color

Why are strawberries red and bananas yellow? How can a soda bottle be green, yet you can still see through it?

If white light is made of all the colors of light, how do things get their color from white light? Why aren't all things white in white light? Good questions! To answer these questions, you need to know how light interacts with matter.

Light and Matter

When light strikes any form of matter, it can interact with the matter in three different ways—the light can be reflected, absorbed, or transmitted.

Reflection happens when light bounces off an object. Reflected light allows you to see things. Absorption is the transfer of light energy to matter. Absorbed light can make things feel warmer. **Transmission** is the passing of light through matter. You see the transmission of light all the time. All of the light that reaches your eyes is transmitted through air. Light can interact with matter in several ways at the same time. Look at **Figure 1**. Light is transmitted, reflected, and absorbed when it strikes the glass in a window.

Figure 1 **Transmission, Reflection, and Absorption**

You can see objects outside because light is **transmitted** through the glass.

You can see the glass and your reflection in it because light is **reflected** off the glass.

The glass feels warm when you touch it because some light is **absorbed** by the glass.

CHAPTER RESOURCES

Chapter Resource File

- Lesson Plan
- Directed Reading A **BASIC**
- Directed Reading B **SPECIAL NEEDS**

Technology

 Transparencies
- Bellringer

Homework —— GENERAL

Writing **Critical Thinking and Writing**
Challenge students to write a paragraph explaining why it is fortunate that some materials and objects transmit light. Why is it fortunate that some things are translucent or opaque? Accept all reasonable answers. **LS Verbal**

Figure 2 Transparent, Translucent, and Opaque

Transparent plastic makes it easy to see what you are having for lunch.

Translucent wax paper makes it a little harder to see exactly what's for lunch.

Opaque aluminum foil makes it impossible to see your lunch without unwrapping it.

Types of Matter

Matter through which visible light is easily transmitted is said to be **transparent.** Air, glass, and water are examples of transparent matter. You can see objects clearly when you view them through transparent matter.

Sometimes, windows in bathrooms are made of frosted glass. If you look through one of these windows, you will see only blurry shapes. You can't see clearly through a frosted window because it is translucent (trans LOO suhnt). **Translucent** matter transmits light but also scatters the light as it passes through the matter. Wax paper is an example of translucent matter.

Matter that does not transmit any light is said to be **opaque** (oh PAYK). You cannot see through opaque objects. Metal, wood, and this book are examples of opaque objects. You can compare transparent, translucent, and opaque matter in **Figure 2.**

✔ **Reading Check** List two examples of translucent objects.
(*See the Appendix for answers to Reading Checks.*)

Colors of Objects

How is an object's color determined? Humans see different wavelengths of light as different colors. For example, humans see long wavelengths as red and short wavelengths as violet. And, some colors, like pink and brown, are seen when certain combinations of wavelengths are present.

The color that an object appears to be is determined by the wavelengths of light that reach your eyes. Light reaches your eyes after being reflected off an object or after being transmitted through an object. When your eyes receive the light, they send signals to your brain. Your brain interprets the signals as colors.

transparent describes matter that allows light to pass through with little interference

translucent describes matter that transmits light but that does not transmit an image

opaque describes an object that is not transparent or translucent

🔵 INCLUSION Strategies

- *Visually Impaired*
- *Learning Disabled*
- *Developmentally Delayed*

Some students have difficulty understanding complicated vocabulary. Use this touch scenario to explain transparent, translucent, and opaque. Place a bowl of water, a ball of modeling clay, and a piece of wood on a table. Ask students to individually put their fingers in the water, poke a hole in the modeling clay, and tap on the wood. Explain that transparent means that they can see through something just like they can put their fingers through the water. Point out that translucent means they can see through something less clearly, just like they can put their fingers through modeling clay less easily. Tell them that opaque means that they cannot see through something at all, just like they cannot put their fingers through the wood. **LS Kinesthetic**

READING STRATEGY — GENERAL

Prediction Guide Before students read the section about colors of opaque objects, ask them whether they would rather sit on a black bench or a white bench if both benches have been exposed to direct sunlight on a hot day. Ask students to explain their answer. Have students read the section, and discuss it with them to make sure they understand why the black bench would be warmer. **LS Verbal** English Language Learners

Answer to Reading Check

When white light shines on a colored opaque object, some of the colors of light are absorbed and some are reflected.

Cultural Awareness — GENERAL

Color Symbolism Colors are used as symbols in many human cultures. For example, red often represents warning or danger, while green can represent safety or safe passage. In the Ukraine, there is a long tradition of egg art—the coloring and decorating of eggs—to express emotions and to send messages. Egg decorators know that red stands for love, green for growth, pink for success, and black for remembrance.

Figure 3 Opaque Objects and Color

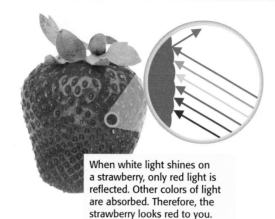

When white light shines on a strawberry, only red light is reflected. Other colors of light are absorbed. Therefore, the strawberry looks red to you.

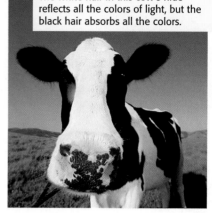

The white hair in this cow's hide reflects all the colors of light, but the black hair absorbs all the colors.

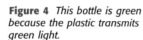

Figure 4 This bottle is green because the plastic transmits green light.

Colors of Opaque Objects

When white light strikes a colored opaque object, some colors of light are absorbed, and some are reflected. Only the light that is reflected reaches your eyes and is detected. So, the colors of light that are reflected by an opaque object determine the color you see. For example, if a sweater reflects blue light and absorbs all other colors, you will see that the sweater is blue. Another example is shown on the left in **Figure 3.**

What colors of light are reflected by the cow shown on the right in **Figure 3**? Remember that white light includes all colors of light. So, white objects—such as the white hair in the cow's hide—appear white because all the colors of light are reflected. On the other hand, black is the absence of color. When light strikes a black object, all the colors are absorbed.

✓ **Reading Check** What happens when white light strikes a colored opaque object?

Colors of Transparent and Translucent Objects

The color of transparent and translucent objects is determined differently than the color of opaque objects. Ordinary window glass is colorless in white light because it transmits all the colors that strike it. But some transparent objects are colored. When you look through colored transparent or translucent objects, you see the color of light that was transmitted through the material. The other colors were absorbed, as shown in **Figure 4.**

CONNECTION ACTIVITY
Earth Science — ADVANCED

Colors of Stars Astronomers use starlight to calculate the temperature of stars. Because of its size and temperature, our sun is listed as a yellow dwarf. Interested students can do research to find out more about how scientists use visible light and other electromagnetic waves to study stars and other objects in space. **LS Verbal**

CHAPTER RESOURCES

Technology

📦 **Transparencies**
• Color Addition

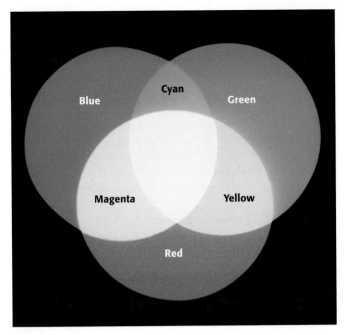

Figure 5 *Primary colors of light—written in white—combine to produce white light. Secondary colors of light—written in black—are the result of two primary colors added together.*

Mixing Colors of Light

In order to get white light, you must combine all colors of light, right? This method is one way of doing it. But you can also get light that appears white by adding just three colors of light together—red, blue, and green. The combination of these three colors is shown in **Figure 5**. In fact, these three colors can be combined in different ratios to produce many colors. Red, blue, and green are called the *primary colors of light.*

Color Addition

When colors of light combine, you see different colors. Combining colors of light is called *color addition.* When two primary colors of light are added together, you see a *secondary color of light.* The secondary colors of light are cyan (blue plus green), magenta (blue plus red), and yellow (red plus green). **Figure 5** shows how secondary colors of light are formed.

Light and Color Television

The colors on a color television are produced by color addition of the primary colors of light. A television screen is made up of groups of tiny red, green, and blue dots. Each dot will glow when the dot is hit by an electron beam. The colors given off by the glowing dots add together to produce all the different colors you see on the screen.

Television Colors

Turn on a color television. Ask an adult to carefully sprinkle a few tiny drops of water onto the television screen. Look closely at the drops of water, and discuss what you see. In your **science journal**, write a description of what you saw.

Q: How many magicians does it take to change a light bulb?

A: Depends on what you want to change it into!

Reteaching — BASIC

Object Colors Show students a variety of opaque and transparent objects. Ask them to identify which colors of light are absorbed and which colors are transmitted by each object. **L⃝S** Visual

Quiz — GENERAL

1. Is the color of a purple car determined the same way as the color of a clear purple bottle? Explain. (No; The purple car reflects purple light and absorbs all other colors; the clear purple bottle transmits purple light and absorbs the other colors.)

2. What color of light will you produce if you mix green light with magenta light? Why? (White; magenta light is a combination of red and blue light, and adding green to it combines all three primary colors to make white light.)

3. How is the color yellow produced on a television screen? (Yellow is produced by mixing the colors emitted by the red and green dots on the screen.)

Alternative Assessment — GENERAL

Concept Mapping Have students make a concept map that shows the three ways that light interacts with matter. The map should include how the color of objects is determined for each interaction. **L⃝S** Verbal

pigment a substance that gives another substance or a mixture its color

Rose-Colored Glasses?

1. Obtain **four plastic filters**—red, blue, yellow, and green.
2. Look through one filter at an object across the room. Describe the object's color.
3. Repeat step 2 with each of the filters.
4. Repeat step 2 with two or three filters together.
5. Why do you think the colors change when you use more than one filter?
6. Write your observations and answers.

Mixing Colors of Pigment

If you have ever tried mixing paints in art class, you know that you can't make white paint by mixing red, blue, and green paint. The difference between mixing paint and mixing light is due to the fact that paint contains pigments.

Pigments and Color

A **pigment** is a material that gives a substance its color by absorbing some colors of light and reflecting others. Almost everything contains pigments. Chlorophyll (KLAWR uh FIL) and melanin (MEL uh nin) are two examples of pigments. Chlorophyll gives plants a green color, and melanin gives your skin its color.

✓ **Reading Check** What is a pigment?

Color Subtraction

Each pigment absorbs at least one color of light. Look at **Figure 6.** When you mix pigments together, more colors of light are absorbed or taken away. So, mixing pigments is called *color subtraction.*

The *primary pigments* are yellow, cyan, and magenta. They can be combined to produce any other color. In fact, every color in this book was produced by using just the primary pigments and black ink. The black ink was used to provide contrast to the images. **Figure 7** shows how the four pigments combine to produce many different colors.

Figure 6 *Primary pigments—written in black—combine to produce black. Secondary pigments—written in white—are the result of the subtraction of two primary pigments.*

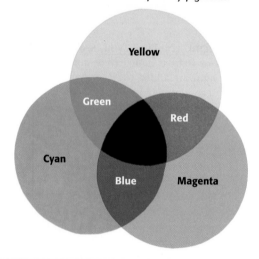

MATERIALS

FOR EACH STUDENT
• plastic filter, 4

Teacher's Notes: Used, but still functioning, filters can be obtained from your school's theater department.

Answers

2–4. The answers will vary, depending on the object chosen by the student and the order in which the different filters are used. Students should notice that a colored object seems to change colors when viewed through different filters or combinations of filters.

5. The colors changed because different colors of light pass through the different filters. If you use more than one filter, fewer colors pass through.

Figure 7 Color Subtraction and Color Printing

The picture of the balloon on the left was made by overlapping yellow ink, cyan ink, magenta ink, and black ink.

Yellow   Cyan   Magenta   Black

SECTION Review

Summary

- Objects are transparent, translucent, or opaque, depending on their ability to transmit light.

- Colors of opaque objects are determined by the color of light that they reflect.

- Colors of translucent and transparent objects are determined by the color of light they transmit.

- White light is a mixture of all colors of light.

- Light combines by color addition. The primary colors of light are red, blue, and green.

- Pigments give objects color. Pigments combine by color subtraction. The primary pigments are magenta, cyan, and yellow.

Using Key Terms

1. Use the following terms in the same sentence: *transmission* and *transparent*.

2. In your own words, write a definition for each of the following terms: *translucent* and *opaque*.

Understanding Key Ideas

3. You can see through a car window because the window is
   a. opaque.   c. transparent.
   b. translucent.   d. transmitted.

4. Name and describe three different ways light interacts with matter.

5. How is the color of an opaque object determined?

6. Describe how the color of a transparent object is determined.

7. What are the primary colors of light, and why are they called *primary colors*?

8. What four colors of ink were used to print this book?

Critical Thinking

9. **Applying Concepts** What happens to the different colors of light when white light shines on an opaque violet object?

10. **Analyzing Ideas** Explain why mixing colors of light is called *color addition* but mixing pigments is called *color subtraction*.

Interpreting Graphics

11. Look at the image below. The red rose was photographed in red light. Explain why the leaves appear black and the petals appear red.

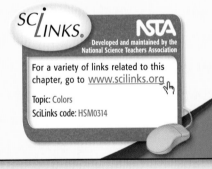

 SCILINKS®
 NSTA
Developed and maintained by the National Science Teachers Association

For a variety of links related to this chapter, go to www.scilinks.org

Topic: Colors
SciLinks code: HSM0314

Answer to Reading Check

A pigment is a material that gives color to a substance by absorbing some colors of light and reflecting others.

CHAPTER RESOURCES

Chapter Resource File

- **Section Quiz** GENERAL
- **Section Review** GENERAL
- **Vocabulary and Section Summary** GENERAL
- **Datasheet for Quick Lab**

Technology

- **Transparencies**
  - Color Subtraction

- **Interactive Explorations CD-ROM**
  - In the Spotlight GENERAL

Mixing Colors

Teacher's Notes

Time Required

One or two 45-minute class periods

Lab Ratings

EASY ———————————→ HARD

Teacher Prep 🧪🧪
Student Set-Up 🧪🧪
Concept Level 🧪🧪
Clean Up 🧪🧪

MATERIALS

Materials listed are for each group of 2–4 students. If a sufficient number of flashlights is not available, consider using the spotlights on the school stage or portable floodlight holders instead. Each group should have a set of watercolors that includes red, blue, and green.

Safety Caution

Students should wear aprons when doing Part B of this lab.

Procedure Notes

For further reinforcement in Part B, students can continue to mix colors to confirm their findings about color subtraction (provided their watercolor sets include more than the three required colors).

Skills Practice Lab

OBJECTIVES

Use flashlights to mix colors of light by color addition.

Use paints to mix colors of pigments by color subtraction.

MATERIALS

Part A
- colored filters, red, green, and blue (1 of each)
- flashlights (3)
- paper, white
- tape, masking

Part B
- cups, small plastic or paper (2)
- paintbrush
- paper, white
- ruler, metric
- tape, masking
- water
- watercolor paints

SAFETY

Mixing Colors

Mix two colors, such as red and green, and you create a new color. Is the new color brighter or darker? Color and brightness depend on the light that reaches your eye. And what reaches your eye depends on whether you are adding colors (mixing colors of light) or subtracting colors (mixing colors of pigments). In this activity, you will do both types of color formation and see the results firsthand!

Part A: Color Addition

Procedure

1. Tape a colored filter over each flashlight lens.

2. In a darkened room, shine the red light on a sheet of white paper. Then, shine the green light next to the red light. You should have two circles of light, one red and one green, next to each other.

3. Move the flashlights so that the circles overlap by half their diameter. What color is formed where the circles overlap? Is the mixed area brighter or darker than the single-color areas? Record your observations.

4. Repeat steps 2 and 3 with the red and blue lights.

5. Now, shine all three lights at the same point on the paper. Record your observations.

Barry Bishop
San Rafael Junior High
Ferron, Utah

CHAPTER RESOURCES

Chapter Resource File

- Datasheet for Chapter Lab
- Lab Notes and Answers

Technology

Classroom Videos
- Lab Video

- What Color of Light Is Best for Green Plants?
- Which Color Is Hottest?

Analyze the Results

1 **Describing Events** In general, when you mixed two colors, was the result brighter or darker than the original colors?

2 **Explaining Events** In step 5, you mixed all three colors. Was the resulting color brighter or darker than when you mixed two colors? Explain your observations in terms of color addition.

Draw Conclusions

3 **Making Predictions** What do you think would happen if you mixed together all the colors of light? Explain your answer.

Part B: Color Subtraction

Procedure

1 Place a piece of masking tape on each cup. Label one cup "Clean" and the other cup "Dirty." Fill each cup about half full with water.

2 Wet the paintbrush thoroughly in the "Clean" cup. Using the watercolor paints, paint a red circle on the white paper. The circle should be approximately 4 cm in diameter.

3 Clean the brush by rinsing it first in the "Dirty" cup and then in the "Clean" cup.

4 Paint a blue circle next to the red circle. Then, paint half the red circle with the blue paint.

5 Examine the three areas: red, blue, and mixed. What color is the mixed area? Does it appear brighter or darker than the red and blue areas? Record your observations.

6 Clean the brush by repeating Step 3. Paint a green circle 4 cm in diameter, and then paint half the blue circle with green paint.

7 Examine the green, blue, and mixed areas. Record your observations.

8 Now add green paint to the mixed red-blue area so that you have an area that is a mixture of red, green, and blue paint. Clean the brush again.

9 Finally, record your observations of this new mixed area.

Analyze the Results

1 **Identifying Patterns** In general, when you mixed two colors, was the result brighter or darker than the original colors?

2 **Analyzing Results** In step 8, you mixed all three colors. Was the result brighter or darker than the result from mixing two colors? Explain what you saw in terms of color subtraction.

Draw Conclusions

3 **Drawing Conclusions** Based on your results, what do you think would happen if you mixed all the colors of paint? Explain your answer.

Part A: Analyze the Results

1. Mixing two colors of light together results in a color that is brighter than the original colors.

2. Mixing three colors of light results in a color that is brighter than the color produced by mixing two colors because more wavelengths are present.

Part A: Draw Conclusions

3. The result would be bright, white light because all the wavelengths of light would be combined.

Part B: Analyze the Results

1. Mixing two colors of paint together results in a color that is darker than the original colors.

2. Mixing three colors of paint results in a color that is darker than the color that results from mixing two colors. Because each color of paint absorbs some light, colors that have been mixed together absorb even more light.

Part B: Draw Conclusions

3. If you mixed all the colors of paint, all colors of light would be absorbed, and a black spot would result.

Disposal Information

Have plenty of paper towels on hand to wipe up water and paint spills. Make sure students clean their brushes thoroughly. Students should use soap and water to clean any watercolor smudges off their lab tables.

Assignment Guide

Section	Questions
1	1, 4, 6, 13, 17
2	12, 14–15
3	3, 8–9, 11, 19, 21–22
4	2, 5, 7, 10, 16, 18, 20

ANSWERS

Using Key Terms

1. Radiation
2. opaque
3. Interference
4. electromagnetic wave
5. transmission

Understanding Key Ideas

6. d
7. c
8. c
9. a
10. c
11. b
12. d
13. c

USING KEY TERMS

Complete each of the following sentences by choosing the correct term from the word bank.

interference radiation
scattering opaque
translucent transmission
electromagnetic electromagnetic
  wave   spectrum

1 _____ is the transfer of energy by electromagnetic waves.

2 This book is a(n) _____ object.

3 _____ is a wave interaction that occurs when two or more waves overlap and combine.

4 Light is a kind of _____ and can therefore travel through matter and space.

5 During _____, light travels through an object.

UNDERSTANDING KEY IDEAS

Multiple Choice

6 Electromagnetic waves transmit
   a. charges.
   b. fields.
   c. matter.
   d. energy.

7 Objects that transmit light easily are
   a. opaque.
   b. translucent.
   c. transparent.
   d. colored.

8 You can see yourself in a mirror because of
   a. absorption.
   b. scattering.
   c. regular reflection.
   d. diffuse reflection.

9 Shadows have blurry edges because of
   a. diffraction.
   b. scattering.
   c. diffuse reflection.
   d. refraction.

10 What color of light is produced when red light is added to green light?
   a. cyan   c. yellow
   b. blue   d. white

11 Prisms produce the colors of the rainbow through
   a. reflection.   c. diffraction.
   b. refraction.   d. interference.

12 Which kind of electromagnetic wave travels fastest in a vacuum?
   a. radio wave
   b. visible light
   c. gamma ray
   d. They all travel at the same speed.

13 Electromagnetic waves are made of
   a. vibrating particles.
   b. vibrating charged particles.
   c. vibrating electric and magnetic fields.
   d. All of the above

Short Answer

14 How are gamma rays used?

15 What are two uses for radio waves?

16 Why is it difficult to see through glass that has frost on it?

Math Skills

17 Calculate the time it takes for light from the sun to reach Mercury. Mercury is 54,900,000 km away from the sun.

CRITICAL THINKING

18 **Concept Mapping** Use the following terms to create a concept map: *light, matter, reflection, absorption,* and *transmission.*

19 **Applying Concepts** A tern is a type of bird that dives underwater to catch fish. When a young tern begins learning to catch fish, the bird is rarely successful. The tern has to learn that when a fish appears to be in a certain place underwater, the fish is actually in a slightly different place. Why does the tern see the fish in the wrong place?

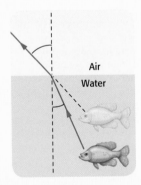

Air
Water

20 **Evaluating Conclusions** Imagine that you are teaching your younger brother about light. You tell him that white light is light of all the colors of the rainbow combined. But your brother says that you are wrong because mixing different colors of paint produces black and not white. Explain why your brother's conclusion is wrong.

21 **Making Inferences** If you look around a parking lot during the summer, you might see sunshades set up in the windshields of cars. How do sunshades help keep the insides of cars cool?

INTERPRETING GRAPHICS

22 Each of the pictures below shows the effects of a wave interaction of light. Identify the interaction involved.

a.

b.

c.

14. Gamma rays are used to treat some forms of cancer and are used to kill harmful bacteria in foods.

15. Two uses for radio waves are the broadcasting of radio signals and the broadcasting of television signals.

16. Frost is translucent, so the light traveling through it is scattered as it passes through.

17. *time = distance ÷ speed*
    time = 54,900,000 km ÷ 300,000 km/s
    time = 183 s (or 3.05 min)

Critical Thinking

18.  An answer to this exercise can be found at the end of this book.

19. The tern sees the fish in the wrong place because light refracts as it passes from the water to the air. This creates an optical illusion for the tern.

20. My brother's conclusion is wrong because he is confusing mixing of light (color addition) with mixing of pigments (color subtraction).

21. Sunshades keep the interior of cars from getting very hot because sunshades reflect light that is transmitted through the glass window of the car. The light reflects back out the window, so the light cannot warm the air or seats inside the car.

Interpreting Graphics

22. **a.** refraction

    b. absorption and scattering (Note: *Reflection* or *diffuse reflection* are also acceptable answers.)

    c. reflection

Standardized Test Preparation

Teacher's Note

To provide practice under more realistic testing conditions, give students 20 minutes to answer all of the questions in this Standardized Test Preparation.

MISCONCEPTION ALERT

Answers to the standardized test preparation can help you identify student misconceptions and misunderstandings.

READING

Passage 1

1. B
2. F

TEST DOCTOR

Question 1: Some students may select answer choice A because the passage states that jaundice can lead to brain damage. However, the passage also states that jaundice is not dangerous if treated quickly. The passage does state that bilirubin is best broken down by bright blue light. Therefore, the correct choice is B.

READING

Read each of the passages below. Then, answer the questions that follow each passage.

Passage 1 Jaundice occurs in some infants when bilirubin—a pigment in healthy red blood cells—builds up in the bloodstream as blood cells break down. This excess bilirubin is deposited in the skin, giving the skin a yellowish hue. Jaundice is not dangerous if treated quickly. If left untreated, it can lead to brain damage.

The excess bilirubin in the skin is best broken down by bright blue light. For this reason, hospitals hang special blue fluorescent lights above the cribs of newborns needing treatment. The blue light is sometimes balanced with light of other colors so that doctors and nurses can be sure the baby is not blue from a lack of oxygen.

1. Which of the following is a fact in the passage?
   - **A** Jaundice is always very dangerous.
   - **B** Bilirubin in the skin of infants can be broken down with bright blue light.
   - **C** Excess bilirubin in the skin gives the skin a bright blue hue.
   - **D** Blue lights can make a baby blue from a lack of oxygen.

2. What is the purpose of this passage?
   - **F** to explain what jaundice is and how it is treated
   - **G** to warn parents about shining blue light on their babies
   - **H** to persuade light bulb manufacturers to make blue light bulbs
   - **I** to explain the purpose of bilirubin in red blood cells

Passage 2 If you have ever looked inside a toaster while toasting a piece of bread, you may have seen thin wires or bars glowing red. The wires give off energy as light when heated to a high temperature. Light produced by hot objects is called *incandescent light*. Most of the lamps in your home probably use incandescent light bulbs.

Sources of incandescent light also release a large amount of <u>thermal</u> energy. Thermal energy is sometimes called *heat energy*. Sometimes, thermal energy from incandescent light is used to cook food or to warm a room. But often this thermal energy is not used for anything. For example, the thermal energy given off by light bulbs is not very useful.

1. What does the word *thermal* mean, based on its use in the passage?
   - **A** light
   - **B** energy
   - **C** heat
   - **D** food

2. What is incandescent light?
   - **F** light used for cooking food
   - **G** light that is red in color
   - **H** light that is not very useful
   - **I** light produced by hot objects

3. Which of the following can be inferred from the passage?
   - **A** Sources of incandescent light are rarely found in an average home.
   - **B** A toaster uses thermal energy to toast bread.
   - **C** Incandescent light from light bulbs is often used to cook food.
   - **D** The thermal energy produced by incandescent light sources is always useful.

Passage 2

1. C
2. I
3. B

TEST DOCTOR

Question 3: The fact that a toaster uses thermal energy to toast bread can be inferred from the passage. The passage states that a toaster is a source of incandescent light and that sources of incandescent light also produce a large amount of thermal energy. The passage also states that thermal energy can be used to cook food. Thus, students should be able to determine that choice B is correct.

The angles of refraction in the table were measured when a beam of light entered the material from air at a 45° angle. Use the table below to answer the questions that follow.

Material and Refraction

Material	Index of refraction	Angle of refraction
Diamond	2.42	17°
Glass	1.52	28°
Quartz	1.46	29°
Water	1.33	32°

1. Which material has the highest index of refraction?

A diamond

B glass

C quartz

D water

2. Which material has the greatest angle of refraction?

F diamond

G glass

H quartz

I water

3. Which of the following statements **best** describes the data in the table?

A The higher the index of refraction, the greater the angle of refraction.

B The higher the index of refraction, the smaller the angle of refraction.

C The greater the angle of refraction, the higher the index of refraction.

D There is no relationship between the index of refraction and the angle of refraction.

4. Which two materials would be the most difficult to separate by observing only their angles of refraction?

F diamond and glass

G glass and quartz

H quartz and water

I water and diamond

Read each question below, and choose the best answer.

1. A square metal plate has an area of 46.3 cm². The length of one side of the plate is between which two values?

A 4 cm and 5 cm

B 5 cm and 6 cm

C 6 cm and 7 cm

D 7 cm and 8 cm

2. A jet was flying over the Gulf of Mexico at an altitude of 2,150 m. Directly below the jet, a submarine was at a depth of −383 m. What was the distance between the jet and the submarine?

F −2,533 m

G −1,767 m

H 1,767 m

I 2,533 m

3. The speed of light in a vacuum is exactly 299,792,458 m/s. Which of the following is a good estimate of the speed of light?

A 3.0×10^{-8} m/s

B 2.0×10^{8} m/s

C 3.0×10^{8} m/s

D 3.0×10^{9} m/s

4. The wavelength of the yellow light produced by a sodium vapor lamp is 0.000000589 m. Which of the following is equal to the wavelength of the sodium lamp's yellow light?

F -5.89×10^{7} m

G 5.89×10^{-9} m

H 5.89×10^{-7} m

I 5.89×10^{7} m

5. Amira purchased a box of light bulbs for $3.81. There are three light bulbs in the box. What is the cost per light bulb?

A $0.79

B $1.06

C $1.27

D $11.43

1. A

2. I

3. B

4. G

TEST DOCTOR

Question 4: To answer this question, students must understand that it is most difficult to separate two materials which have similar angles of refraction. According to the table, the angles of refraction of quartz and glass differ by only one degree. Therefore, G is the correct choice.

MATH

1. C

2. I

3. C

4. H

5. C

TEST DOCTOR

Question 1: To answer this question, students must find the square root of 46.3 cm² to find the length of one side of the square. The square root of 46.3 cm² is approximately 6.8 cm. Therefore, the correct choice is C.

Standardized Test Preparation

CHAPTER RESOURCES

Chapter Resource File

 • Standardized Test Preparation GENERAL

State Resources

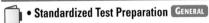 For specific resources for your state, visit **go.hrw.com** and type in the keyword **HSMSTR**.

Weird Science

Background

Luciferase assays have a number of relevant uses in medicine as well. Bacterial infections can be deadly. Researchers have developed luciferase assays to measure bacteria in the urine and blood. Antibiotics, the medicines used to treat bacterial infections, are not always effective. Luciferase tests can be used to evaluate the effectiveness of antibiotic therapy on particular patients.

Science, Technology, and Society

Background

Microwave radiation is very different from radioactive radiation. The radiation that microwave ovens use is called *nonionizing radiation*. This is not to be confused with the ionizing radiation of X rays, gamma rays, and cosmic rays that can cause damage to living cells (including genetic mutations and tissue damage) and alter the molecular structure of matter. Examples of nonionizing radiation are radio waves, infrared radiation, and visible light. These waves have lower frequencies and less energy than ionizing radiation.

Science in Action

Weird Science

Fireflies Light the Way

Just as beams of light from lighthouses warn boats of approaching danger, the light of an unlikely source—fireflies—is being used by scientists to warn food inspectors of bacterial contamination.

Fireflies use an enzyme called *luciferase* to make light. Scientists have taken the gene from fireflies that tells cells how to make luciferase. They put this gene into a virus that preys on bacteria. The virus is not harmful to humans and can be mixed into meat. When the virus infects bacteria in the meat, the virus transfers the gene into the genes of the bacteria. The bacteria then produce luciferase and glow! So, if a food inspector sees glowing meat, the inspector knows that the meat is contaminated with bacteria.

Science, Technology, and Society

It's a Heat Wave

In 1946, Percy Spencer visited a laboratory belonging to Raytheon—the company he worked for. When he stood near a device called a *magnetron,* he noticed that a candy bar in his pocket melted. Spencer hypothesized that the microwaves produced by the magnetron caused the candy bar to warm up and melt. To test his hypothesis, Spencer put a bag of popcorn kernels next to the magnetron. The microwaves heated the kernels, causing them to pop! Spencer's simple experiment showed that microwaves could heat foods quickly. Spencer's discovery eventually led to the development of the microwave oven—an appliance found in many kitchens today.

Social Studies

WRITING SKILL Many cultures have myths to explain certain natural phenomena. Read some of these myths. Then, write your own myth titled "How Fireflies Got Their Fire."

Math ACTiViTY

Popcorn pops when the inside of the kernel reaches a temperature of about 175°C. Convert this temperature to degrees Fahrenheit.

Answer to Social Studies Activity

Accept all reasonable answers. Encourage students to be creative when writing their myths. Invite volunteers to read their myths to the class.

Answer to Math Activity

The equation for converting degrees Celsius to degrees Fahrenheit is $°F = (9/5 \times °C) + 32$. So, $(9/5 \times 175°C) + 32 = 347°F$.

Albert Einstein

A Light Pioneer When Albert Einstein was 15 years old, he asked himself, "What would the world look like if I were speeding along on a motorcycle at the speed of light?" For many years afterward, he would think about this question and about the very nature of light, time, space, and matter. He even questioned the ideas of Isaac Newton, which had been widely accepted for 200 years. Einstein was bold. And he was able to see the universe in a totally new way.

In 1905, Einstein published a paper on the nature of light. He knew from the earlier experiments of others that light was a wavelike phenomenon. But he theorized that light could also travel as particles. Scientists did not readily accept Einstein's particle theory of light. Even 10 years later, the American physicist Robert Millikan, who proved that the particle theory of light was true, was reluctant to believe his own experimental results. Einstein's theory helped pave the way for television, computers, and other important technologies. The theory also earned Einstein a Nobel Prize in physics in 1921.

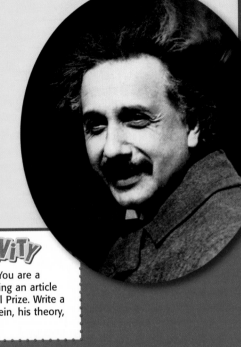

Language Arts ACTIVITY

WRITING SKILL Imagine that it is 1921. You are a newspaper reporter writing an article about Albert Einstein and his Nobel Prize. Write a one-page article about Albert Einstein, his theory, and the award he won.

To learn more about these Science in Action topics, visit go.hrw.com and type in the keyword HP5LGTF.

Current Science

Check out Current Science® articles related to this chapter by visiting go.hrw.com. Just type in the keyword HP5CS22.

People in Science

Research ———— ADVANCED

The theory described in this feature is often referred to as Einstein's theory concerning the photoelectric effect. Four important theories were advanced by Einstein in 1905. Only one of those theories, Einstein's hypothesis concerning the nature of light, is described in this article. You may wish to assign students the task of finding out about the other three theories Einstein advanced in 1905.

(One theory was about the dimension of molecules. It earned Einstein a doctorate from the University of Zurich in mid-1905. A second theory was about the motion of particles that are randomly distributed in a fluid. These ideas were based on the work of a Scottish botanist named Robert Brown. This theory is sometimes referred to as the theory of Brownian motion. The third theory is now known as the special theory of relativity. This theory led to the famous equation $E = mc^2$. It replaced many of Isaac Newton's assertions about the nature of the universe. The equation $E = mc^2$ shows that the energy content of matter (E) equals mass (m) times the speed of light (c) squared. This concept was so radical that it would not be accepted or proven for many years.)

Answers to Language Arts Activity
Accept all reasonable responses. Students may need to research Einstein and his theory before writing their articles.

Compression guide:
To shorten instruction
because of time limitations,
omit Section 3.

OBJECTIVES	LABS, DEMONSTRATIONS, AND ACTIVITIES	TECHNOLOGY RESOURCES
PACING • 90 min pp. 96–103 **Chapter Opener**	SE **Start-up Activity,** p. 97 GENERAL	OSP **Parent Letter** ■ GENERAL CD **Student Edition on CD-ROM** CD **Guided Reading Audio CD** ■ TR **Chapter Starter Transparency*** VID **Brain Food Video Quiz**
Section 1 Mirrors and Lenses • Use ray diagrams to show how light is reflected or refracted. • Compare plane mirrors, concave mirrors, and convex mirrors. • Use ray diagrams to show how mirrors form images. • Describe the images formed by concave and convex lenses.	TE **Activity** Light and Mirrors, p. 98 GENERAL TE **Activity** Mirror Height, p. 99 ADVANCED TE **Activity** Spoon Reflections, p. 100 GENERAL SE **School-to-Home Activity** Car Mirrors, p. 101 GENERAL SE **Skills Practice Lab** Images from Convex Lenses, p. 116 GENERAL CRF **Datasheet for Chapter Lab*** SE **Skills Practice Lab** Mirror Images, p. 132 GENERAL CRF **Datasheet for LabBook*** LB **Inquiry Labs** Eye Spy* ADVANCED	TR **Lesson Plans*** TR **Bellringer Transparency*** TR **How Images Are Formed in Plane Mirrors*** TR **The Optical Axis, Focal Point, and Focal Length*** TR **How Concave Mirrors Form Images*** TR **How Light Passes Through Lenses*** SE **Internet Activity,** p. 100 GENERAL CRF **SciLinks Activity*** GENERAL VID **Lab Videos for Physical Science**
PACING • 45 min pp. 104–107 **Section 2 Light and Sight** • Identify the parts of the human eye, and describe their functions. • Describe three common vision problems. • Describe surgical eye correction.	TE **Activity** Eye See You!, p. 104 GENERAL TE **Activity** Nearsighted and Farsighted Projector, p. 105 BASIC TE **Connection Activity** Language Arts, p. 105 ADVANCED SE **Connection to Biology** Color Deficiency and Genes, p. 106 GENERAL TE **Connection Activity** Math, p. 106 GENERAL LB **Whiz-Bang Demonstrations** Light Humor* GENERAL	CRF **Lesson Plans*** TR **Bellringer Transparency*** TR **How Your Eyes Work*** CD **Science Tutor**
PACING • 45 min pp. 108–115 **Section 3 Light and Technology** • Describe three optical instruments. • Explain what laser light is, and identify uses for lasers. • Describe how optical fibers work. • Explain polarized light. • Explain how radio waves and microwaves are used in four types of communication technology.	TE **Demonstration** Optical Instruments, p. 108 GENERAL TE **Connection Activity** Math, p. 108 ADVANCED TE **Activity** Microscope History, p. 109 BASIC TE **Activity** Laser Model, p. 110 BASIC TE **Demonstration** Polarizing Filter and Liquid Crystals, p. 112 GENERAL SE **Quick Lab** Blackout!, p. 113 GENERAL CRF **Datasheet for Quick Lab*** SE **Connection to Social Studies** Navigation, p. 114 GENERAL SE **Science in Action** Math, Social Studies, and Language Arts Activities, pp. 122–123 GENERAL LB **Labs You Can Eat** Fiber Optic Fun* GENERAL LB **EcoLabs & Field Activities** Photon Drive* ADVANCED LB **Long-Term Projects & Research Ideas** Island Vacation* ADVANCED	CRF **Lesson Plans*** TR **Bellringer Transparency*** TR **How a Camera Works*** TR **How Refracting and Reflecting Telescopes Work*** TR *LINK TO LIFE SCIENCE* **Compound Light Microscope*** CD **Science Tutor**

PACING • 90 min

CHAPTER REVIEW, ASSESSMENT, AND STANDARDIZED TEST PREPARATION

CRF **Vocabulary Activity*** GENERAL
SE **Chapter Review,** pp. 118–119 GENERAL
CRF **Chapter Review*** ■ GENERAL
CRF **Chapter Tests A*** ■ GENERAL, **B*** ADVANCED, **C*** SPECIAL NEEDS
SE **Standardized Test Preparation,** pp. 120–121 GENERAL
CRF **Standardized Test Preparation*** GENERAL
CRF **Performance-Based Assessment*** GENERAL
OSP **Test Generator** GENERAL
CRF **Test Item Listing*** GENERAL

Online and Technology Resources

Visit **go.hrw.com** for a variety of free resources related to this textbook. Enter the keyword **HP5LOW.**

Holt Online Learning

Students can access interactive problem-solving help and active visual concept development with the *Holt Science and Technology* Online Edition available at **www.hrw.com.**

 Guided Reading Audio CD
Also in Spanish

A direct reading of each chapter for auditory learners, reluctant readers, and Spanish-speaking students.

 Science Tutor
CD-ROM

Excellent for remediation and test practice.

SKILLS DEVELOPMENT RESOURCES	SECTION REVIEW AND ASSESSMENT	CORRELATIONS
SE **Pre-Reading Activity,** p. 96 GENERAL OSP **Science Puzzlers, Twisters & Teasers** GENERAL		National Science Education Standards UCP 2, 3; SAI 1, 2; SPSP 5
CRF **Directed Reading A*** ■ BASIC, **B*** SPECIAL NEEDS CRF **Vocabulary and Section Summary*** ■ GENERAL SE **Reading Strategy** Reading Organizer, p. 98 GENERAL TE **Reading Strategy** Mnemonics, p. 100 GENERAL TE **Inclusion Strategies,** p. 101 CRF **Reinforcement Worksheet** Mirror, Mirror* BASIC	SE **Reading Checks,** pp. 99, 100, 102 GENERAL TE **Homework,** p. 99 GENERAL TE **Homework,** p. 100 GENERAL TE **Reteaching,** p. 102 BASIC TE **Quiz,** p. 102 GENERAL TE **Alternative Assessment,** p. 102 GENERAL SE **Section Review,*** p. 103 ■ GENERAL CRF **Section Quiz*** ■ GENERAL	UCP 2; PS 3c; *Chapter Lab:* UCP 3; SAI 1; PS 3c; *LabBook:* PS 3c
CRF **Directed Reading A*** ■ BASIC, **B*** SPECIAL NEEDS CRF **Vocabulary and Section Summary*** ■ GENERAL SE **Reading Strategy** Reading Organizer, p. 104 GENERAL	SE **Reading Checks,** pp. 105, 106 GENERAL TE **Homework,** p. 105 GENERAL TE **Reteaching,** p. 106 BASIC TE **Quiz,** p. 106 GENERAL TE **Alternative Assessment,** p. 106 GENERAL SE **Section Review,*** p. 107 ■ GENERAL CRF **Section Quiz*** ■ GENERAL	UCP 2; SPSP 5; PS 3c
CRF **Directed Reading A*** ■ BASIC, **B*** SPECIAL NEEDS CRF **Vocabulary and Section Summary*** ■ GENERAL SE **Reading Strategy** Prediction Guide, p. 108 GENERAL SE **Math Practice** Microscope Magnification, p. 109 GENERAL TE **Inclusion Strategies,** p. 111 CRF **Critical Thinking** Light That Heals* ADVANCED	SE **Reading Checks,** pp. 109, 111, 113, 114 GENERAL TE **Homework,** p. 110 GENERAL TE **Homework,** p. 113 ADVANCED TE **Reteaching,** p. 114 BASIC TE **Quiz,** p. 114 GENERAL TE **Alternative Assessment,** p. 114 GENERAL SE **Section Review,*** p. 115 ■ GENERAL CRF **Section Quiz*** ■ GENERAL	UCP 5; SAI 1; ST 2; SPSP 5; PS 3c

One-Stop Planner® CD-ROM

This CD-ROM package includes:
- Lab Materials QuickList Software
- Holt Calendar Planner
- Customizable Lesson Plans
- Printable Worksheets
- ExamView® Test Generator
- Interactive Teacher Edition
- Holt PuzzlePro® Resources
- Holt PowerPoint® Resources

SCLINKS
NSTA

www.scilinks.org

Maintained by the **National Science Teachers Association.** See Chapter Enrichment pages for a complete list of topics.

Current Science®

Check out **Current Science** articles and activities by visiting the HRW Web site at **go.hrw.com.** Just type in the keyword **HP5CS23T.**

 Classroom Videos

- **Lab Videos** demonstrate the chapter lab.
- **Brain Food Video Quizzes** help students review the chapter material.
- **CNN Videos** bring science into your students' daily life.

Visual Resources

CHAPTER STARTER TRANSPARENCY

Light and Our World — CHAPTER STARTER

Imagine . . .

When *Apollo 11* astronauts Neil Armstrong and Edwin "Buzz" Aldrin completed the first moonwalk in 1969, they left more than footprints on the moon. Armstrong and Aldrin also left a small panel called a retroreflector that consists of 100 light-reflecting cubes.

Ever since, scientists on Earth have been aiming light beams at the retro-reflector—more than 370,000 km away. The beams come from a laser, which is a device that produces intense light of only one wavelength. The laser light is directed through a telescope, reflected off a mirror inside the telescope, and aimed toward the retroreflector. By the time the laser beam reaches the retro-reflector, it has spread to cover an area more than 6 km wide.

The retroreflector, which is about 46 cm on a side, reflects a portion of the laser beam back to Earth. Scientists record the time it takes for the light to return to the laboratory. Using these data, scientists have been able to measure the distance between Earth and the moon to within 3 cm!

Researchers around the world continue to bounce laser beams off the retroreflector and continue to learn more about the moon. For example, scientists have determined that the moon is slowly moving away from Earth and that the length of an Earth day changes slightly over the course of a year.

In this chapter you will learn more about light sources, such as lasers, and the mirrors and lenses that are used to build telescopes. All these things are not just useful to scientists; they are a part of your world!

BELLRINGER TRANSPARENCIES

Light and Our World — BELLRINGER TRANSPARENCY

Section: Mirrors and Lenses

What is the difference between a mirror and a lens? What is the difference between a convex mirror and a concave mirror? Can you think of one common use for lenses?

Record your responses in your **science journal**.

Section: Light and Sight

What do you think a person who has colorblindness sees? If you have colorblindness, describe what you see. What difficulties would a person who has colorblindness have?

Record your answers in your **science journal**.

TEACHING TRANSPARENCIES

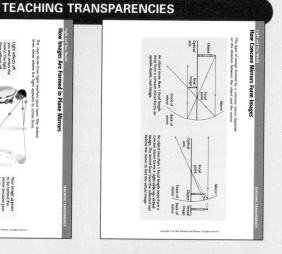

How Concave Mirrors Form Images

How Images Are Formed in Plane Mirrors

TEACHING TRANSPARENCIES

How Your Eyes Work

How a Camera Works

How Light Passes Through Lenses

When light rays pass through a convex lens, the rays are refracted toward each other.

When light rays pass through a concave lens, the rays are refracted away from each other.

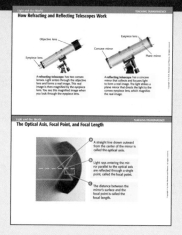

How Refracting and Reflecting Telescopes Work

The Optical Axis, Focal Point, and Focal Length

Compound Light Microscope

Ocular lens
Body tube
Revolving nosepiece
Objective lens
Stage clip
Stage
Diaphragm
Light

Fine-adjustment knob
Coarse-adjustment knob
Arm

LINK TO LIFE SCIENCE

Chapter: The World of Life Science

CONCEPT MAPPING TRANSPARENCY

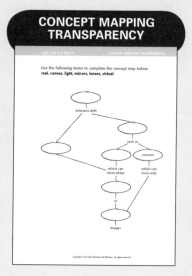

Light and Our World — CONCEPT MAPPING TRANSPARENCY

Use the following terms to complete the concept map below:
real, convex, light, mirrors, lenses, virtual

Planning Resources

LESSON PLANS

Lesson Plan SAMPLE

Section: Waves

Pacing

Regular Schedule: with lab(s):2 days without lab(s):if days
Block Schedule: with lab(s):1 1/2 days without lab(s):1 day

Objectives

1. Relate the seven properties of life to a living organism.
2. Describe seven themes that can help you to organize what you learn about biology.
3. Identify the tiny structures that make up all living organisms.
4. Differentiate between reproduction and heredity and between metabolism and homeostasis.

National Science Education Standards Covered

LSInter1:Cells have particular structures that underlie their functions.
LSMat1: Most cell functions involve chemical reactions.
LSBeh1:Cells store and use information to guide their functions.
UCP1:Cell functions are regulated.
SE1: Cells can differentiate from complete multicellular organisms.
PS1: Species evolve over time.
ESS1: The great diversity of organisms is the result of more than 3.5 billion years of evolution.
ESS2: Natural selection and its evolutionary consequences provide a scientific explanation for the fossil record of ancient life forms as well as for the striking molecular similarities observed among the diverse species of living organisms.
ST1: The millions of different species of plants, animals, and microorganisms that live on Earth today are related through common ancestors.
ST2: The energy for life primarily comes from the sun.
SPS1: The complexity and organization of organisms accommodates the need for obtaining, transforming, transporting, releasing, and eliminating the matter and energy used to sustain the organism.
SPS%: As matter and energy flows through different levels of organization of living systems—cells, organs, communities—and through living systems and the physical environment, chemical elements are recombined in different ways.
HNS1: Organisms have behavioral responses to internal change and to external stimuli.

PARENT LETTER

SAMPLE

Dear Parent,

Your son's or daughter's science class will soon begin exploring the chapter entitled "The World of Physical Science." In this chapter, students will learn about how the scientific method applies to the world of physical science and the role of physical science in the world. By the end of the chapter, students should demonstrate a clear understanding of the chapter's main ideas and be able to discuss the following topics:

1. physical science is the study of energy and matter (Section 1)
2. the role of physical science in the world around them (Section 1)
3. careers that rely on physical science (Section 1)
4. the steps used in the scientific method (Section 2)
5. examples of technology (Section 2)
6. how the scientific method is used to answer questions and solve problems (Section 2)
7. how your knowledge of science changes over time (Section 2)
8. how models represent real objects or systems (Section 3)
9. examples of different ways models are used in science (Section 3)
10. the importance of the International System of Units (Section 4)
11. the appropriate units to use for particular measurements (Section 4)
12. how area and density are derived quantities (Section 4)

Questions to Ask Along the Way

You can help your son or daughter learn about these topics by asking interesting questions such as the following:

• What are some surprising careers that use physical science?
• What is a characteristic of a good hypothesis?
• When is it a good idea to use a model?
• Why do Americans measure things in terms of inches and yards instead of centimeters and meters ?

ALSO IN SPANISH

TEST ITEM LISTING

TEST ITEM LISTING
The World of Science SAMPLE

MULTIPLE CHOICE

1. A limitation of models is that
   a. they are large enough to see.
   b. they do not act exactly like the things that they model.
   c. they are smaller than the things that they model.
   d. they model unfamiliar things.
   Answer: B   Difficulty: 1   Section: 3   Objective: 2

2. The length 10 m is equal to
   a. 100 cm.   c. 10,000 mm.
   b. 1,000 cm.   d. Both (b) and (c)
   Answer: D   Difficulty: 1   Section: 3   Objective: 2

3. To be valid, a hypothesis must be
   a. testable.   c. made into a law.
   b. supported by evidence. d. Both (a) and (b)
   Answer: B   Difficulty: 1   Section: 2   Objective: 2   1

4. The statement "Sheila has a stain on her shirt" is an example of a(n)
   a. law.   c. observation.
   b. hypothesis.   d. prediction.
   Answer: B   Difficulty: 1   Section: 2   Objective: 2

5. A hypothesis is often developed out of
   a. observations.   c. laws.
   b. experiments.   d. Both (a) and (b)
   Answer: B   Difficulty: 1   Section: 3   Objective: 2

6. How many milliliters are in 3.5 kL?
   a. 3,500 mL.   c. 3,500,000 mL.
   b. 0.0035 mL.   d. 3,500 mL.
   Answer: B   Difficulty: 1   Section: 3   Objective: 2

7. A map of Seattle is an example of a
   a. law.   c. model.
   b. theory.   d. unit.
   Answer: B   Difficulty: 1   Section: 3   Objective: 2

8. A lab has the safety icons shown below. These icons mean that you should wear
   a. only safety goggles.   c. safety goggles and a lab apron.
   b. only a lab apron.   d. safety goggles, a lab apron, and gloves.
   Answer: B   Difficulty: 1   Section: 1   Objective: 2

9. The law of conservation of mass says the lot of mass before a chemical change is
   a. more than the total mass after the change.
   b. less than the total mass after the change.
   c. the same as the total mass after the change.
   d. not the same as the total mass after the change.
   Answer: B   Difficulty: 1   Section: 3   Objective: 2

10. In which of the following areas might you find a geochemist at work?
    a. studying the chemistry of rocks   c. studying fishes
    b. studying forestry   d. studying the atmosphere
    Answer: A   Difficulty: 1   Section: 1   Objective: 2

One-Stop Planner® CD-ROM

This CD-ROM includes all of the resources shown here and the following time-saving tools:

• **Lab Materials QuickList Software**
• **Customizable lesson plans**
• **Holt Calendar Planner**
• **The powerful ExamView® Test Generator**

Meeting Individual Needs

DIRECTED READING A
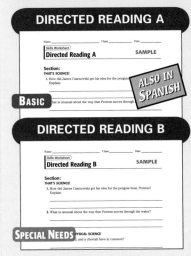

Skills Worksheet
Directed Reading A — SAMPLE

Section:
THAT'S SCIENCE!
1. How did James Czarnowski get his idea for the penguin boat
Explain.

ALSO IN SPANISH

BASIC

that is unusual about the way that Proteus moves through

DIRECTED READING B

Name _____ Class _____ Date _____
Skills Worksheet
Directed Reading B — SAMPLE

Section:
THAT'S SCIENCE!
1. How did James Czarnowski get his idea for the penguin boat, Proteus?
Explain.

2. What is unusual about the way that Proteus moves through the water?

SPECIAL NEEDS PHYSICAL SCIENCE
a, and a cheetah have in common?

VOCABULARY ACTIVITY
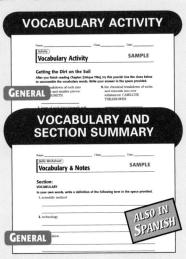

Activity
Vocabulary Activity — SAMPLE

Getting the Dirt on the Soil
After you finish reading Chapter [Unique Title], try this puzzle! Use the clues below to unscramble the vocabulary words. Write your answer in the space provided.

GENERAL

9. the chemical breakdown of rocks

VOCABULARY AND SECTION SUMMARY

Name _____ Class _____ Date _____
Skills Worksheet
Vocabulary & Notes — SAMPLE

Section:
VOCABULARY
In your own words, write a definition of the following term in the space provided.
1. scientific method

2. technology

ALSO IN SPANISH

GENERAL

REINFORCEMENT
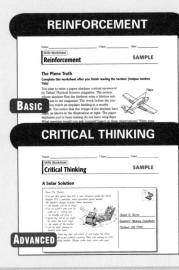

Skills Worksheet
Reinforcement — SAMPLE

The Plane Truth
Complete this worksheet after you finish reading the Section: [Unique Section Title]

You plan to enter a paper airplane contest sponsored by Talkin' Physical Science magazine. The person whose airplane flies the farthest wins a lifetime subscription to the magazine! The week before the contest you watch an airplane landing at a nearby airport. You notice that the wings of the airplane have flaps, as shown in the illustration at right. The paper airplanes you've been testing do not have wing flaps.
What question would you ask yourself based on these observations? Write your

Flaps

BASIC

CRITICAL THINKING

Skills Worksheet
Critical Thinking — SAMPLE

A Solar Solution

ADVANCED

SCILINKS ACTIVITY

Activity
SciLinks Activity — SAMPLE

MARINE ECOSYSTEMS
Go to www.scilinks.org to find links related to marine ecosystems, type in the keyword HL5800. Then, use the links to answer the questions about marine ecosys-

Topic: Reproductive System
Irregulation
Irregulation
SciLinks code: HL5800

percentage of the Earth's surface is covered by water?

GENERAL

SCIENCE PUZZLERS, TWISTERS & TEASERS

CHAPTER
23 SCIENCE PUZZLERS, TWISTERS & TEASERS
Light and Our World

Riddle-Eye-O
1. "See" if you can answer the riddles below. Write each item in the space provided.
a. In bright light I hardly show, but dimness causes me to grow.
b. Brown, green, hazel, or blue, I provide the eye its hue.
c. I focus light to the back of the eye, and change shape for things far and nigh.
d. I'm clear on my role to protect the eye and refract the light as it passes by.
of the eye is where I am found;
mage forms on me, upside down.

2. Each clue below will lead to one or two short words. Combine

GENERAL

Labs and Activities

ECOLABS & FIELD ACTIVITIES

Name _____ Date _____ Class _____
EcoLab
23 STUDENT WORKSHEET
Photon Drive

DISCOVERY LAB

The people have spoken! They are tired of parking meters, ridiculous parking fees, and awful lunch-time traffic! Buses solve half the problem, but what if you need to get from one side of downtown to the other in a hurry? Buses still get stuck in the noon-time traffic and they still pollute.

What a great opportunity for your mass transit company! You have recently decided to merge with a small solar energy company to develop a solar-powered tram. This car, called the Arrow Transport, would travel separate routes from one side of town to the other with stops in between. The system would reduce downtown traffic, pollution, and parking hassles.

Your company has put in a bid, but your company still compete with several other providers to get the contract. The city council will choose only the fastest and straightest-traveling vehicle for this project! You and your group submit their meeting tomorrow to create your prototype. Good luck!

MATERIALS
• tongue depressor
• scissors
• plastic drinking straw
• bamboo skewer
• metric ruler
• 3 film canister lids
• 2 pencil erasers
• wood glue
• 1.5 V–3.0 V motor
• rubber bands
• hook-and-loop adhesive tape
• sheet of corrugated cardboard
• wire coils
• modeling clay
• 2 insulated wires with alligator clips
• aluminum foil

Objective
Construct a solar car to explore the use of an alternative fuel.

Procedure
1. Cut the tongue depressor to three-fourths its original length. This will serve as the body of the car.
2. Cut the straw in half. Pull the loose fibers from the bamboo skewer to reduce friction, and place it inside the straw. Cut the bamboo skewer so that it is 2.5 cm longer than the straw piece. This will serve as the axle of the car.
3. Make tiny holes in the center of two of the film canister lids for the rear wheels. Pull the erasers off two new pencils. Connect a lid to each end of the bamboo skewer, and secure the lids in place with the erasers.
4. Glue the axle to one end of the tongue depressor so that it forms a T, as shown on the next page.
5. Mount the motor sideways on the other end of the tongue depressor, as shown on the next page. The motor shaft should be mounted so that it is parallel to the bamboo skewer at the opposite end.

ADVANCED

WHIZ-BANG DEMONSTRATIONS
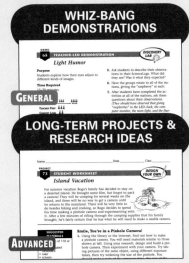

DEMO
6S TEACHER-LED DEMONSTRATION
Light Humor

DISCOVERY LAB

Purpose
Students explore how their eyes adjust to different kinds of images.

Time Required
15–30 minutes

5. Ask students to describe their observations in their ScienceLogs. What did they see? Was it what they expected?
6. Have the groups rotate to all of the stations, giving the "raspberry" at each.
7. After students have completed the activities at all of the stations, ask them questions about their observations. (They should have observed that giving "raspberries" to the LED clock, the computer monitor, the neon light, and the fluo-

TEACHER PREP
CONCEPT LEVEL

GENERAL

LONG-TERM PROJECTS & RESEARCH IDEAS

Name _____ Date _____ Class _____
PROJECT
73 STUDENT WORKSHEET
Island Vacation

DESIGN YOUR OWN

For summer vacation Bogo's family has decided to stay on a deserted island. He brought some film, but forgot to pack a camera! They will be camping for several weeks on the island, and there will be no way to get a camera until he returns to the mainland. There will be very little to do besides hiking and cooking, so Bogo decides to spend his time making a pinhole camera and experimenting with it. After a few minutes of rifling through the camping supplies that his family brought, he's fairly certain that he has what he will need to make a usable camera.

SUGGESTED MATERIALS
of 110 or
• ruler
• scissors

Smile, You're in a Pinhole Camera!
1. Using the library or the Internet, find out how to make a pinhole camera. You will need materials similar to those shown at left. Using your research, design and build a pinhole camera. Then experiment with your camera. Try taking pictures of the same object, using different exposure times, then try widening the size of the pinhole. You

ADVANCED

INQUIRY LABS

LAB
23 STUDENT WORKSHEET
Eye Spy

DISCOVERY LAB

Surveillance 101: Introduction to Investigation
The life of a private investigator is quite extraordinary. In this six-week course, you will learn the basics of a private investigator's work: 30 Easy Meals for a Stake-Out, A Micro-Camera for Every Event, 201 Easy Disguises, and Simple Surveillance Techniques—Watch Without Being Watched.
Your first assignment requires you to build a periscope for discreet surveillance. You are not allowed to purchase the periscope; a good private investigator is resourceful enough to build one. So get ready to make a scope and sneak a peek!

Ask a Question
How can the principles of reflection be used to build a periscope?

• pocket-sized mirror
• sheet of graph paper

ADVANCED

LABS YOU CAN EAT

LAB
25 STUDENT WORKSHEET
Fiber-Optic Fun

MAKING MODELS

Have you ever seen telephone workers installing new telephone cables along the highway? The cables are probably not made of copper wire, which can conduct electrical energy, but of fine, glass fibers that carry light. As amazing as it seems, telephone companies use optical fibers to send telephone conversations in the form of tiny pulses of light.
Telephone companies make use of the process of total internal reflection to send a signal from weakening too rapidly over long distances or from leaking out of the fiber as the signal is transmitted down the line.

In this activity, your teacher will demonstrate how fiber optics are used to trans-

GENERAL

DATASHEETS FOR QUICK LABS

TEACHER RESOURCE PAGE
Quick Lab
Reaction to Stress — DATASHEET FOR QUICK LAB — SAMPLE

Background
The graph below illustrates changes that occur in the membrane potential of a neuron during an action potential. Use the graph to answer the following questions. Refer to Figure 3 on page.

DATASHEETS FOR CHAPTER LABS

TEACHER RESOURCE PAGE
Skills Practice Lab
Using Scientific Methods — DATASHEET FOR CHAPTER LAB — SAMPLE

Teacher's Notes
TIME REQUIRED
One 45-minute class period.

DATASHEETS FOR LABBOOK

TEACHER RESOURCE PAGE
Skills Practice Lab
Does It All Add Up? — DATASHEET FOR LABBOOK LAB — SAMPLE

Teacher's Notes
TIME REQUIRED
One 45-minute class period.

Review and Assessments

SECTION QUIZ
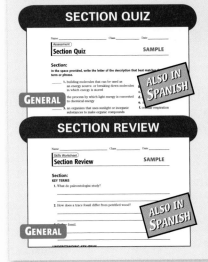

Name _____ Class _____ Date _____
Assessment
Section Quiz — SAMPLE

Section:
In the space provided, write the letter of the description that best matches the term or phrase.

____ 1. building molecules that can be used as an energy source, or breaking down molecules in which energy is stored
____ 2. the process by which light energy is converted to chemical energy
____ 3. an organism that uses sunlight or inorganic substances to make organic compounds

ALSO IN SPANISH

GENERAL

SECTION REVIEW

Name _____ Class _____ Date _____
Skills Worksheet
Section Review — SAMPLE

Section:
KEY TERMS
1. What do paleontologist study?

2. How does a trace fossil differ from petrified wood?

fossil.

UNDERSTANDING KEY IDEAS

ALSO IN SPANISH

GENERAL

CHAPTER REVIEW
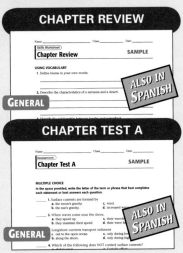

Name _____ Class _____ Date _____
Skills Worksheet
Chapter Review — SAMPLE

USING VOCABULARY
1. Define biome in your own words.

2. Describe the characteristics of a savanna and a desert.

ALSO IN SPANISH

GENERAL

CHAPTER TEST A

Name _____ Class _____ Date _____
Assessment
Chapter Test A — SAMPLE

MULTIPLE CHOICE
In the space provided, write the letter of the term or phrase that best completes each statement or best answers each question.

____ 1. Surface currents are formed by
  a. the moon's gravity.   c. wind.
  b. the sun's gravity.   d. increased water density.
____ 2. When waves come near the shore,
  a. they speed up.   c. their wavel
  b. they maintain their speed.   d. their wave
  Longshore currents transport sediment
  a. out to the open ocean.   c. only during low tide.
  b. along the shore.   d. only during high
____ 4. Which of the following does NOT control surface currents?

ALSO IN SPANISH

GENERAL

CHAPTER TEST B
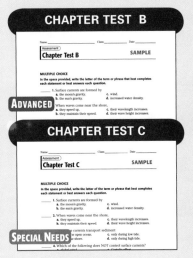

Assessment
Chapter Test B — SAMPLE

MULTIPLE CHOICE
In the space provided, write the letter of the term or phrase that best completes each statement or best answers each question.

____ 1. Surface currents are formed by
  a. the moon's gravity.   c. wind.
  b. the sun's gravity.   d. increased water density.
____ When waves come near the shore,
  a. they speed up.   c. their wavelength increases.
  b. they maintain their speed.   d. their wave height increases.

ADVANCED

CHAPTER TEST C

Assessment
Chapter Test C — SAMPLE

MULTIPLE CHOICE
In the space provided, write the letter of the term or phrase that best completes each statement or best answers each question.

____ 1. Surface currents are formed by
  a. the moon's gravity.   c. wind.
  b. the sun's gravity.   d. increased water density.
____ 2. When waves come near the shore,
  a. they speed up.   c. their wavelength increases.
  b. they maintain their speed.   d. their wave
  currents transport sediment
  open ocean.   c. only during low tide.
                                  d. only during high tide.
____ 4. Which of the following does NOT control surface currents?

SPECIAL NEEDS

STANDARDIZED TEST PREPARATION
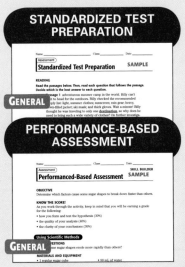

Assessment
Standardized Test Preparation — SAMPLE

READING
Read the passages below. Then, read each question that follows the passage. Decide which is the best answer to each question.

Passage 1 adventurous summer camp in the world. Billy can't wait to head for the outdoors. Billy checked the recommended supply list: light, summer clothes, suntscreen, rain gear, heavy, heavy-filled jacket; ski mask and thick gloves. Wait a minute! Billy thought he was traveling to only one destination, so why does he need to bring such a wide variety of clothes? On further investi-

GENERAL

PERFORMANCE-BASED ASSESSMENT

Assessment
Performance-Based Assessment — SKILL BUILDER — SAMPLE

OBJECTIVE
Determine which factors cause some sugar shapes to break down faster than others.

KNOW THE SCORE!
As you work through the activity, keep in mind that you will be earning a grade for the following:
• how you form and test the hypothesis (30%)
• the quality of your analysis (40%)
• the clarity of your conclusions (30%)

Using Scientific Methods
QUESTIONS

sugar shapes erode more rapidly than others?

MATERIALS AND EQUIPMENT
• 1 regular sugar cube   • 60 mL of water

GENERAL

This Chapter Enrichment provides relevant and interesting information to expand and enhance your presentation of the chapter material.

Section 1

Mirrors and Lenses

History of Mirrors

- Natural mirrors made of obsidian were used in Turkey 7,500 years ago. Bronze mirrors were used in Egypt as early as 3500 to 3000 BCE. Later, polished mirrors of copper, brass, bronze, tin, and silver were used.

- Metal mirrors were luxury items because it was difficult to make a flat, highly polished metal surface that would reflect light well enough to form images.

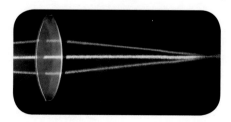

- The Venetians found a way to use polished silver to make mirrors in the 1200s, but the silvering process used on mirrors today was founded by German chemist Justus von Liebig (1803–1873) in 1835.

Is That a Fact!

- ◆ The Keck and Keck II telescopes, on Mauna Kea in Hawaii, are 10 m in diameter. They are the largest reflecting telescopes in the world today. Each uses 36 mirror segments fitted together so seamlessly that they act as one large mirror. The segments are realigned about 1,000 times a second by computers to counteract the effects of gravity and other distortions.

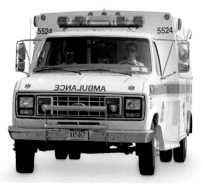

Early Uses of Eyeglasses

- No one knows for sure who invented eyeglasses. Some claim that eyeglasses were in use in China during the 13th century. Others say that eyeglasses originated in Arabia in the 11th century. However, it is known that eyeglasses came into use in Italy around the 1280s.

Section 2

Light and Sight

Correcting Vision

- Adapting lasers for medical use enables doctors to correct previously untreatable eye problems. Laser surgery can repair the retina and is used in cataract surgery, cornea replacement, and vision-correction surgery.

Stereoscopic Vision

- Having two eyes allows humans to judge depth, distance, and speed effectively. Each eye receives a slightly different view of the same object, and the brain combines these views to give a three-dimensional interpretation. People who suffer from strabismus, a defect in which the eyes are not used together, often have difficulty judging distances.

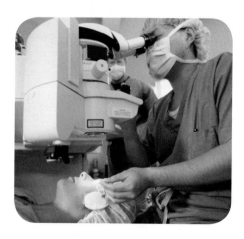

Restored Vision

- A landmark study published in 2003 focused on the experiences of Michael May, a man whose vision was restored after about 40 years of total blindness. May lost his sight when he was 3 years old. When his vision was first restored, he could not interpret what he was seeing. Two years after regaining his sight, May could detect motion and color but still had difficulties identifying objects and people and interpreting facial expressions.

- May's experiences led scientists to suggest that people develop some visual processes, such as detecting motion, very early in life. But complex processing of visual signals by the brain, such as that needed to recognize people, develops later in life.

Is That a Fact!

- The human eye is so sensitive to light that in the dark the eye can see a lighted candle 1.6 km away.

- The human body needs vitamin A to produce the light-sensitive protein rhodopsin in the rods of the eye. Both the structure of the rods and the rhodopsin enable the eyes to see in dim light.

Section 3

Light and Technology

Holograms

- The term *hologram* is a compound word derived from two Greek words. *Holo* is Greek for "whole," and *gram* is Greek for "message."

- In 1948, Dennis Gabor (1900–1979) invented the process for making holograms. The first holograms were of poor quality because a good source of coherent light was not available. When lasers were perfected in the 1960s, holography surged in popularity.

- If a hologram is made using a source that has a short wavelength, such as ultraviolet light, and then is viewed in visible light, which has longer wavelengths, the image produced appears greatly magnified. The amount of magnification is proportional to the ratio of the increase in wavelengths.

Endoscopes

- Endoscopes are medical tools that have been used since 1958 to look inside the body by using optical fibers. The long, thin tubes carry light without distortion to and from the area being observed. Observation is made through an eyepiece. Optical fibers can be thinner than human hair.

Is That a Fact!

- In optical fibers, total internal reflection depends on the shallow angle (critical angle) at which the light waves reflect off the walls of the optical fiber. The critical angle when water acts as the prism is 49°; for crown glass (a type of optical glass), it is 41.1°.

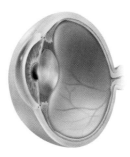

Geocaching

- Geocaching is a relatively new, treasure-hunt game that uses GPS receivers. A geocache is a container with a logbook and several small items inside. Geocaches are placed all over the world by geocaching players.

- To play, a person uses the Internet to look up latitude and longitude coordinates and hints for finding a geocache. Then, he or she enters the latitude and longitude into a GPS receiver and attempts to find the geocache.

- Once a player finds a geocache, he or she writes in the logbook. The player may write when the geocache was found or whatever he or she wishes to share. A player may take an item from a geocache but should leave another item in its place.

SciLINKS

NSTA
Developed and maintained by the
National Science Teachers Association

SciLinks is maintained by the National Science Teachers Association to provide you and your students with interesting, up-to-date links that will enrich your classroom presentation of the chapter.

Visit www.scilinks.org and enter the SciLinks code for more information about the topic listed.

Topic: Mirrors
SciLinks code: HSM0970

Topic: The Eye
SciLinks code: HSM0560

Topic: Lenses
SciLinks code: HSM0868

Topic: Lasers
SciLinks code: HSM0853

Overview

In this chapter, students will learn how mirrors and lenses form images. Students will also study the way the human eye works and the causes of certain vision problems. Finally, students will learn about several kinds of light technology including optical instruments, lasers, and polarized light.

Assessing Prior Knowledge

Students should be familiar with the following topics:

• electromagnetic waves

• reflection and refraction

• the law of reflection

Identifying Misconceptions

As students learn the material in this chapter, some of them may think that a person who has color deficiency, or colorblindness, can see only in black and white. Explain to students that people who have color deficiency can see some colors, but they do not see all colors in the same way that a person who has normal vision does. For example, a person who has red-green color deficiency may see the colors red and green as yellow or tan. True colorblindness, in which a person can see only in black and white, is very rare.

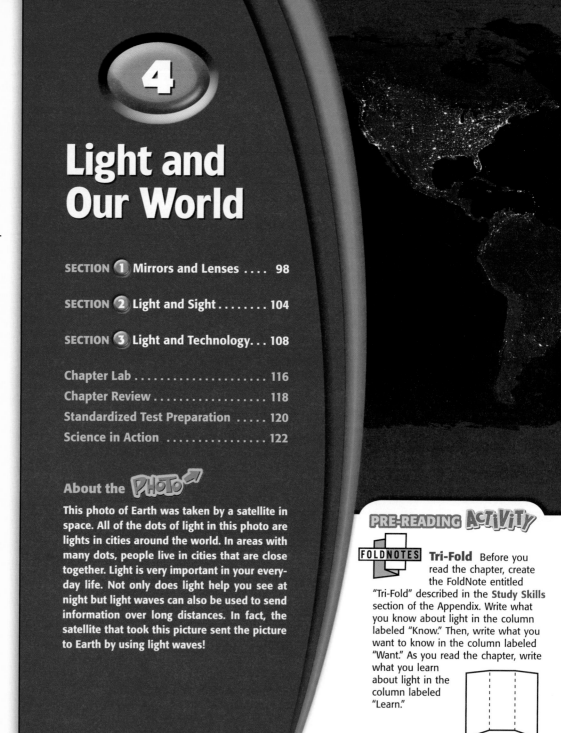

Light and Our World

About the PHOTO

This photo of Earth was taken by a satellite in space. All of the dots of light in this photo are lights in cities around the world. In areas with many dots, people live in cities that are close together. Light is very important in your everyday life. Not only does light help you see at night but light waves can also be used to send information over long distances. In fact, the satellite that took this picture sent the picture to Earth by using light waves!

PRE-READING ACTIVITY

FOLDNOTES **Tri-Fold** Before you read the chapter, create the FoldNote entitled "Tri-Fold" described in the **Study Skills** section of the Appendix. Write what you know about light in the column labeled "Know." Then, write what you want to know in the column labeled "Want." As you read the chapter, write what you learn about light in the column labeled "Learn."

Standards Correlations

National Science Education Standards

The following codes indicate the National Science Education Standards that correlate to this chapter. The full text of the standards is at the front of the book.

Chapter Opener
UCP 2, 3; SAI 1, 2; SPSP 5

Section 1 Mirrors and Lenses
UCP 2; PS 3c; *LabBook*: PS 3c

Section 2 Light and Sight
UCP 2; SPSP 5; PS 3c

Section 3 Light and Technology
UCP 5; SAI 1; ST 2; SPSP 5; PS 3c

Chapter Lab
UCP 3; SAI 1; PS 3c

Chapter Review
SAI 1; PS 3c

Science in Action
ST 1, 2; SPSP 5; HNS 1, 3

START-UP ACTIVITY

MATERIALS

FOR EACH GROUP
- clay, modeling
- glass, colored
- mirror, flat
- paper, graph, 1 cm squares
- pen
- ruler, metric

Safety Caution: Have students wear safety goggles while doing this activity. Also caution students to handle the mirrors and pieces of colored glass very carefully. Tape the edges of the glass with masking tape.

Answers to Procedure

2. The pen should be four squares behind the mirror. The pen should appear farther behind the mirror than before.

3. The image in the glass should be the same size and same distance behind the glass as the image was in the mirror.

4. The length of the sides of both squares should be identical.

Answers to Analysis

1. In general, the distance from an object to a plane mirror and the distance from the mirror to the image is the same.

2. In general, the size of an object and the size of its image in a plane mirror are identical.

START-UP ACTIVITY

Mirror, Mirror

In this activity, you will study images formed by flat, or plane, mirrors.

Procedure

1. Tape a sheet of **graph paper** on your desk. Stand a **flat mirror** in the middle of the paper. Hold the mirror in place with pieces of **modeling clay.**

2. Place a **pen** four squares in front of the mirror. How many squares behind the mirror is the image of the pen? Move the pen farther away from the mirror. How did the image change?

3. Replace the mirror with **colored glass.** Look at the image of the pen in the glass. Compare the image in the glass with the one in the mirror.

4. Draw a square on the graph paper in front of the glass. Then, look through the glass, and trace the image of the square on the paper behind the glass. Using a **metric ruler,** measure and compare the two squares.

Analysis

1. How does the distance from an object to a plane mirror compare with the apparent distance from the mirror to the object's image behind the mirror?

2. Images formed in the colored glass are similar to images formed in a plane mirror. In general, how does the size of an object compare with that of its image in a plane mirror?

Imagine . . .

When Apollo 11 astronauts Neil Armstrong and Edwin "Buzz" Aldrin completed the first moonwalk in 1969, they left more than footprints on the moon. Armstrong and Aldrin also left a small panel called a retroreflector that consists of 100 light-reflecting cubes.

Ever since, scientists on Earth have been aiming light beams at the retroreflector—more than 370,000 km away. The beams come from a laser, which is a device that produces intense light of only one wavelength. The laser light is directed through a telescope, reflected off a mirror inside the telescope, and aimed toward the retroreflector. By the time

data, scientists have been able to measure the distance between Earth and the moon to within 3 cm!

Researchers around the world continue to bounce laser beams off the retroreflector and continue to learn more about the moon. For example, scientists have determined that the moon is slowly moving away from Earth and that the

Chapter Starter Transparency
Use this transparency to help students begin thinking about mirrors and lenses.

CHAPTER RESOURCES

Technology

Transparencies
- Chapter Starter Transparency

READING SKILLS

 Student Edition on CD-ROM

Guided Reading Audio CD
- English or Spanish

 **Classroom Videos**
- Brain Food Video Quiz

Workbooks

 Science Puzzlers, Twisters & Teasers
- Light and Our World **GENERAL**

Focus

Overview

In this section, students learn how mirrors and lenses form images and how ray diagrams are used to determine where the images are. This section covers plane mirrors, concave mirrors, convex mirrors, convex lenses, and concave lenses.

 **Bellringer**

Ask students the following questions:

- What is the difference between a mirror and a lens?

- What is the difference between a convex mirror and a concave mirror?

- What is one common use for lenses?

Motivate

 ACTiViTY ──────── **GENERAL**

Light and Mirrors Give each group of students two small mirrors and a flashlight. Tell students to experiment with the mirrors and flashlights to find out how light travels and what mirrors do to light. Dim the classroom lights so students can see the beams of light better. Students should learn that light travels in a straight line and that light continues to travel in a straight line after reflecting off a mirror. **Kinesthetic**

READING WARM-UP

Objectives

- Use ray diagrams to show how light is reflected or refracted.
- Compare plane mirrors, concave mirrors, and convex mirrors.
- Use ray diagrams to show how mirrors form images.
- Describe the images formed by concave and convex lenses.

Terms to Learn

plane mirror   lens
concave mirror   convex lens
convex mirror   concave lens

READING STRATEGY

Reading Organizer As you read this section, make a concept map by using the terms above.

Mirrors and Lenses

When walking by an ambulance, you notice that the letters on the front of the ambulance look strange. Some letters are backward, and they don't seem to spell a word!

Look at **Figure 1.** The letters spell the word *ambulance* when viewed in a mirror. Images in mirrors are reversed left to right. The word *ambulance* is spelled backward so that people driving cars can read it when they see an ambulance in their rearview mirrors. To understand how images are formed in mirrors, you must first learn how to use rays to trace the path of light waves.

Rays and the Path of Light Waves

Light waves are electromagnetic waves. Light waves travel from their source in all directions. If you could trace the path of one light wave as it travels away from a light source, you would find that the path is a straight line. Because light waves travel in straight lines, you can use an arrow called a *ray* to show the path and the direction of a light wave.

Rays and Reflected and Refracted Light

Rays help to show the path of a light wave after it bounces or bends. Light waves that bounce off an object are reflected. Light waves that bend when passing from one medium to another are refracted. So, rays in ray diagrams show changes in the direction light travels after being reflected by mirrors or refracted by lenses.

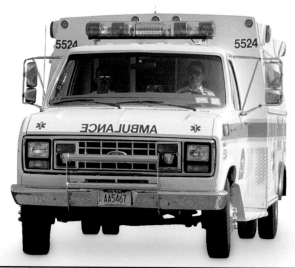

Figure 1 *If you hold this photo up to the mirror in your bathroom, you will see the word* AMBULANCE.

CHAPTER RESOURCES

Chapter Resource File

- **Lesson Plan**
- **Directed Reading A** BASIC
- **Directed Reading B** SPECIAL NEEDS

Technology

Transparencies
- Bellringer
- How Images Are Formed in Plane Mirrors

SCIENTISTS AT ODDS

Particles or Waves? Sir Isaac Newton (1642–1727) did not accept the theory of his colleague Robert Hooke (1635–1703) that light is a wave. Newton believed that white light was composed of particles, or "corpuscles." Newton knew that if light were a wave, it should bend around corners. When Newton could not prove that light bends around corners, he disagreed with Hooke.

Mirrors and Reflection of Light

Have you ever looked at your reflection in a metal spoon? The spoon is like a mirror but not like a bathroom mirror! If you look on one side of the spoon, your face is upside down. But on the other side, your face is right side up. Why? Read on to find out!

The shape of a mirror affects the way light reflects from it. So, the image you see in your bathroom mirror differs from the image you see in a spoon. Mirrors are classified by their shape. Three shapes of mirrors are plane, concave, and convex.

Plane Mirrors

Most mirrors, such as the one in your bathroom, are plane mirrors. A **plane mirror** is a mirror that has a flat surface. When you look in a plane mirror, your reflection is right side up. The image is also the same size as you are. Images in plane mirrors are reversed left to right, as shown in **Figure 2.**

In a plane mirror, your image appears to be the same distance behind the mirror as you are in front of it. Why does your image seem to be behind the mirror? When light reflects off the mirror, your brain thinks the reflected light travels in a straight line from behind the mirror. The ray diagram in **Figure 3** explains how light travels when you look into a mirror. The image formed by a plane mirror is a virtual image. A *virtual image* is an image through which light does not travel.

✓ **Reading Check** What is a virtual image? (*See the Appendix for answers to Reading Checks.*)

Figure 2 *Rearview mirrors in cars are plane mirrors. This mirror shows the reflection of the front of the ambulance shown in **Figure 1.***

plane mirror a mirror that has a flat surface

Figure 3 — How Images Are Formed in Plane Mirrors

The rays show how light reaches your eyes. The dotted lines show where the light appears to come from.

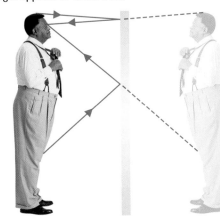

Light reflects off you and strikes the mirror. The light then reflects off the mirror at an angle equal to the angle at which the light hit the mirror. Some of the reflected light enters your eyes.

Your image appears to be behind the mirror because your brain assumes that the light rays that enter your eyes travel in a straight line from an object to your eyes.

Homework — GENERAL

Build a Periscope Have students make periscopes by using materials of their choice. Each student should draw a diagram explaining how his or her periscope works. Ask students to think of ways in which periscopes can be used. **LS** Kinesthetic

Answer to Reading Check

A virtual image is an image through which light does not travel.

Teach

ACTIVITY — ADVANCED

Mirror Height Have students use what they know about reflection to explain why a plane mirror must be at least half a person's height for the person to see his or her full image in the mirror. Have students use diagrams or mirrors in their explanations. (The angle of incidence equals the angle of reflection, so a person can see the top of his or her head by looking at a point on the mirror halfway between his or her eyes and the top of his or her head. A person can see his or her feet by looking at a point on the mirror halfway between his or her eyes and feet. Together, these two images add up to half the person's height. [Note: The mirror should be hung so that the top of the mirror is midway between the top of the person's head and his or her eyes.]) **LS** Logical/Visual

CONNECTION to Environmental Science — GENERAL

Reflective Windows Up to 40% of the summertime heat that builds up in a house is a direct result of sunlight that shines through the windows. Special coatings can be applied to windows that reflect up to 80% of the incoming sunlight. These coatings are partially reflective, much like a plane mirror, but they transmit enough light for a person to see through the window. Adding a reflective coating to windows that receive direct sunlight can reduce the amount of energy needed to cool a home.

Mnemonics Students are learning about concave and convex mirrors and lenses and about the types of images each produces. Have students create mnemonic devices to help them recall the differences between concave and convex. (Sample answer: Concave curves inward because it has caved in.) **LS** Verbal

ACTIVITY — GENERAL

Spoon Reflections Ask students to predict what type of image they will see when they look at themselves in the inside of a spoon. Ask them to draw the image they predict they will see. Then, give students spoons, and have students look at their reflection. Have them draw what they actually see. Then, have students move the spoon closer and closer to their eye and describe what they see. Repeat the activity with the back of the spoon. English Language Learners **LS** Kinesthetic

Answer to Reading Check

A concave mirror can be used to make a powerful beam of light by putting a light source at the focal point of the mirror.

Figure 4 *Concave mirrors are curved like the inside of a spoon. The image formed by a concave mirror depends on the optical axis, focal point, and focal length of the mirror.*

concave mirror a mirror that is curved inward like the inside of a spoon

convex mirror a mirror that is curved outward like the back of a spoon

INTERNET ACTIVITY

For another activity related to this chapter, go to **go.hrw.com** and type in the keyword **HP5LOWW**.

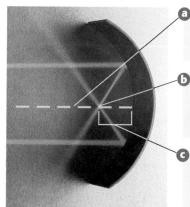

a A straight line drawn outward from the center of the mirror is called the **optical axis**.

b Light rays entering the mirror parallel to the optical axis are reflected through a single point, called the **focal point**.

c The distance between the mirror's surface and the focal point is called the **focal length**.

Concave Mirrors

A mirror that is curved inward is called a **concave mirror**. The images formed by concave mirrors differ from the images formed by plane mirrors. The image formed by a concave mirror depends on three things: the optical axis, focal point, and focal length of the mirror. **Figure 4** explains these terms.

You have already learned that plane mirrors can form only virtual images. Concave mirrors also form virtual images. But they can form real images, too. A *real image* is an image through which light passes. A real image can be projected onto a screen, but a virtual image cannot.

Concave Mirrors and Ray Diagrams

To find out what kind of image a concave mirror forms, you can make a ray diagram. Draw two rays from the top of the object to the mirror. Then, draw rays reflecting from the surface of the mirror. If the reflected rays cross in front of the mirror, a real image is formed. If the reflected rays do not cross in front of the mirror, extend the reflected rays in straight lines behind the mirror. Those lines will cross to show where a virtual image is formed. Study **Figure 5** to better understand ray diagrams.

If an object is placed at the focal point of a concave mirror, no image will form. All rays that pass through the focal point on their way to the mirror will reflect parallel to the optical axis. The rays will never cross in front of or behind the mirror. If you put a light source at the focal point of a concave mirror, light will reflect outward in a powerful beam. So, concave mirrors are used in car headlights and flashlights.

✓ Reading Check How can a concave mirror be used to make a powerful beam of light?

MISCONCEPTION ALERT

Making Beams of Light Be sure your students understand that to produce a beam of light by placing a candle at the focal point of a concave mirror, the flame of the candle (not the base of the candle) should be at the focal point.

Homework — GENERAL

 Lighthouses Some lighthouses use a strong light source placed at or near the focal point of a concave mirror. Ask students to write a story about a lighthouse. Students may write about a sailor who uses the light from a lighthouse while at sea or about the life of a lighthouse operator. **LS** Verbal

Figure 5 *The type of image formed by a concave mirror depends on the distance between the object and the mirror.*

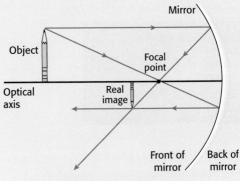 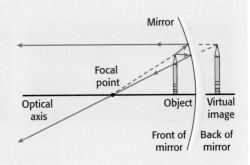

An object more than 1 focal length away from a concave mirror forms an **upside-down, real image.**

An object less than 1 focal length away from a concave mirror forms a **right-side-up, virtual image.** The dotted lines trace the reflected rays behind the mirror to find the virtual image.

Convex Mirrors

If you look at your reflection in the back of a spoon, you will notice that your image is right side up and small. The back of a spoon is a convex mirror. A **convex mirror** is a mirror that curves outward. **Figure 6** shows how an image is formed by a convex mirror. The reflected rays do not cross in front of a convex mirror. So, the reflected rays are extended behind the mirror to find the virtual image. All images formed by convex mirrors are virtual, right side up, and smaller than the original object. Convex mirrors are useful because they make images of large areas. So, convex mirrors are often used for security in stores and factories. Convex mirrors are also used as side mirrors on cars and trucks.

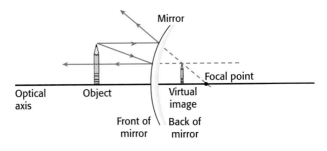

Figure 6 *All images formed by convex mirrors are formed behind the mirror. Therefore, all images formed by convex mirrors are virtual.*

SCHOOL to HOME

Car Mirrors

Sit in the passenger side of a car. Ask an adult at home to stand one car-length behind the car. Look at the adult's reflection in the passenger side mirror. Then, look at the adult's reflection in the rearview mirror. Make a table comparing the two mirrors and the images you saw in each mirror.

ACTIVITY

Is That a Fact!

Most people know that if they look at the inside, or concave side, of a spoon, they will see their upside-down reflection. However, the inside of a spoon, like all concave mirrors, forms an image that is right side up if an object is placed less than one focal length away from the spoon. If you hold a spoon very close to one eye, you will see a right-side-up reflection of your eye.

Reteaching ——— BASIC

Using Convex Lenses Give each pair of students a magnifying lens. Tell students to use the lens to look at an object. First, tell students to use the lens to see an image that is larger than the object. Have students draw diagrams showing the relative positions of their eye, the magnifying lens, and the object. Then, challenge students to use the lens to see an image that is smaller than the object. Ask them to draw a diagram for this situation, too. **LS Kinesthetic**

Quiz ——— GENERAL

1. What is the difference between a real image and a virtual image? (Light actually passes through a real image, and a real image can be projected onto a screen. Neither is true of a virtual image.)

2. What is the difference between a mirror and a lens? (A mirror reflects light; a lens refracts light.)

3. List three objects that contain lenses. (Sample answers: cameras, telescopes, movie projectors, eyeglasses, magnifying lenses, microscopes)

Alternative Assessment ——— GENERAL

Concept Mapping Have students make a concept map explaining the properties of lenses and the images they form. **LS Verbal/Visual**

Figure 7 How Light Passes Through Lenses

When light rays pass through a **convex lens**, the rays are refracted toward each other.

When light rays pass through a **concave lens**, the rays are refracted away from each other.

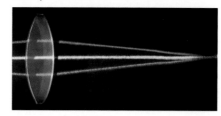
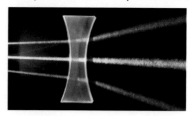

lens a transparent object that refracts light waves such that they converge or diverge to create an image

convex lens a lens that is thicker in the middle than at the edges

concave lens a lens that is thinner in the middle than at the edges

Lenses and Refraction of Light

What do cameras, telescopes, and movie projectors have in common? They all use lenses to create images. A **lens** is a transparent object that forms an image by refracting, or bending, light. Lenses are classified by their shape. Two kinds of lenses, convex and concave, are shown in **Figure 7**. The yellow beams in **Figure 7** show that light rays that pass through the center of any lens are not refracted. Like mirrors, lenses have a focal point and an optical axis.

Convex Lenses

A **convex lens** is a lens that is thicker in the middle than at the edges. Convex lenses form different kinds of images. The ways in which two of these kinds of images are formed are shown in **Figure 8**. In addition, a convex lens can form a real image that is larger than the object if the object is between 1 and 2 focal lengths away from the lens. Convex lenses have many uses. For example, magnifying lenses and camera lenses are convex lenses. And convex lenses are sometimes used in eyeglasses.

✓ **Reading Check** What is a convex lens?

Figure 8 *The distance between an object and a convex lens determines the size and the kind of image formed.*

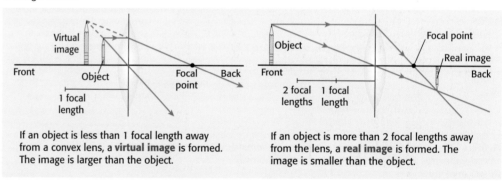

If an object is less than 1 focal length away from a convex lens, a **virtual image** is formed. The image is larger than the object.

If an object is more than 2 focal lengths away from the lens, a **real image** is formed. The image is smaller than the object.

Answer to Reading Check

A convex lens is thicker in the middle than it is at the edges.

Concave Lenses

A **concave lens** is a lens that is thinner in the middle than at the edges. Light rays entering a concave lens parallel to the optical axis always bend away from each other and appear to come from a focal point in front of the lens. The rays never meet. So, concave lenses never form a real image. Instead, they form virtual images, as shown in **Figure 9.** Concave lenses are sometimes combined with other lenses in telescopes. The combination of lenses produces clearer images of distant objects. Concave lenses are also used in microscopes and eyeglasses.

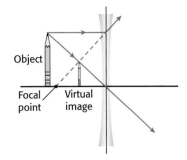

Figure 9 *Concave lenses form virtual images. The image is smaller than the object.*

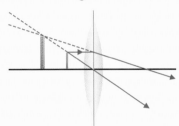

SECTION Review

Summary

- Rays are arrows that show the path of a single light wave.
- Ray diagrams can be used to find where images are formed by mirrors and lenses.
- Plane mirrors and convex mirrors produce virtual images. Concave mirrors produce both real images and virtual images.
- Convex lenses produce both real images and virtual images. Concave lenses produce only virtual images.

Using Key Terms

For each pair of terms, explain how the meanings of the terms differ.

1. *convex mirror* and *concave mirror*

2. *convex lens* and *concave lens*

Understanding Key Ideas

3. Which of the following can form real images?
   a. a plane mirror
   b. a convex mirror
   c. a convex lens
   d. a concave lens

4. Explain how you can use a ray diagram to determine if a real image or a virtual image is formed by a mirror.

5. Compare the images formed by plane mirrors, concave mirrors, and convex mirrors.

6. Describe the images that can be formed by convex lenses.

7. Explain why a concave lens cannot form a real image.

Critical Thinking

8. **Applying Concepts** Why is an image right side up on the back of a spoon but upside down on the inside of a spoon?

9. **Making Inferences** Teachers sometimes use overhead projectors to show transparencies on a screen. What type of lens does an overhead projector use?

Interpreting Graphics

10. Look at the ray diagram below. Identify the type of lens and the kind of image that is formed.

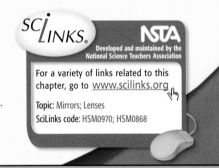

SCILINKS **NSTA**

Developed and maintained by the National Science Teachers Association

For a variety of links related to this chapter, go to www.scilinks.org

Topic: Mirrors; Lenses
SciLinks code: HSM0970; HSM0868

Answers to Section Review

1. Sample answer: A convex mirror is curved outward like the back of a spoon, and a concave mirror is curved inward like the inside of a spoon.

2. Sample answer: A convex lens is thicker in the middle than at the edges, and a concave lens is thinner in the middle than at the edges.

3. c

4. A real image is formed if the rays in a ray diagram cross in front of the mirror. A real image forms where the rays cross. A virtual image is formed if the reflected rays cross when they are extended behind the mirror. A virtual image forms where the extended rays cross.

5. Plane mirrors and convex mirrors can form only virtual images. Concave mirrors can form both virtual images and real images.

6. Convex lenses can form virtual images that are larger than the object, real images that are larger than the object, and real images that are smaller than the object.

7. A concave lens cannot form a real image because rays passing through a concave lens bend outward and never meet.

8. The inside of a spoon is a concave mirror and forms a real, upside down image. The back of a spoon is a convex mirror and forms a virtual image that is right side up. (Note: The inside of a spoon can also form an image that is right side up, but the object must be less than one focal length away from the spoon. Students may hold a small object close to the inside of a spoon to see this effect.)

9. An overhead projector uses a convex lens. Convex lenses can form real images, and concave lenses cannot. Only real images can be projected onto a screen.

10. The lens is a convex lens. The image formed is a virtual image.

SECTION

2

Focus

Overview

This section identifies the parts of the human eye and explains how they function. It also describes some common vision problems and the ways they can be corrected.

Bellringer

Ask students the following questions:

• What do you think a person who has colorblindness sees? If you have colorblindness, describe what you do see.

• What difficulties would a person who has colorblindness have?

Motivate

ACTIVITY ———— GENERAL

Eye See You! Have pairs of students observe one another's eyes. Tell students to carefully watch each other's pupils as you turn the lights off and on. Turn the lights off and on a few times, waiting several seconds between each flip of the switch. Ask students why the pupils change size.

LS Visual

English Language Learners

READING WARM-UP

Objectives

● Identify the parts of the human eye, and describe their functions.

● Describe three common vision problems.

● Describe surgical eye correction.

Terms to Learn

nearsightedness
farsightedness

READING STRATEGY

Reading Organizer As you read this section, make a flowchart of how the eye works.

Light and Sight

When you look around, you can see objects both near and far. You can also see the different colors of the objects.

You see objects that produce their own light because the light is detected by your eyes. You see all other objects because light reflected from the objects enters your eyes. But how do your eyes work, and what causes people to have vision problems?

How You Detect Light

Visible light is the part of the electromagnetic spectrum that can be detected by your eyes. Your eye gathers light to form the images that you see. The steps of this process are shown in **Figure 1.** Muscles around the lens change the thickness of the lens so that objects at different distances can be seen in focus. The light that forms the real image is detected by receptors in the retina called *rods* and *cones*. Rods can detect very dim light. Cones detect colors in bright light.

Figure 1 How Your Eyes Work

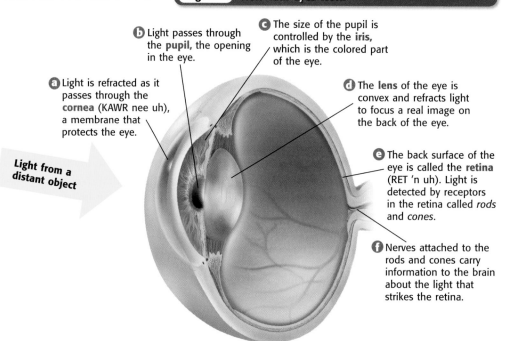

b Light passes through the **pupil,** the opening in the eye.

c The size of the pupil is controlled by the **iris,** which is the colored part of the eye.

a Light is refracted as it passes through the **cornea** (KAWR nee uh), a membrane that protects the eye.

d The **lens** of the eye is convex and refracts light to focus a real image on the back of the eye.

Light from a distant object

e The back surface of the eye is called the **retina** (RET 'n uh). Light is detected by receptors in the retina called *rods* and *cones.*

f Nerves attached to the rods and cones carry information to the brain about the light that strikes the retina.

CHAPTER RESOURCES

Chapter Resource File

• Lesson Plan
• Directed Reading A **BASIC**
• Directed Reading B **SPECIAL NEEDS**

Technology

Transparencies
• Bellringer
• How Your Eyes Work

CONNECTION to Life Science ——— GENERAL

Cataracts In some people, the lens of the eye gets cloudy. This cloudy lens is called a *cataract.* Eye surgeons correct cataracts by making a tiny incision in the eye and inserting a slender instrument that uses sound waves to break up the cloudy lens. The surgeon then removes the pieces. Once the cataract is removed, a plastic implant lens is inserted inside the eye. The surgery takes about an hour, and the patient can go home the same day.

Figure 2 Correcting Nearsightedness and Farsightedness

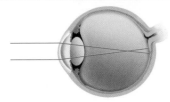

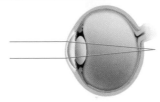

Nearsightedness happens when the eye is too long, which causes the lens to focus light in front of the retina.

Farsightedness happens when the eye is too short, which causes the lens to focus light behind the retina.

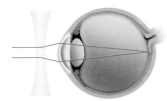

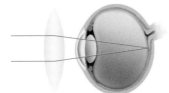

A **concave lens** placed in front of a nearsighted eye refracts the light outward. The lens in the eye can then focus the light on the retina.

A **convex lens** placed in front of a farsighted eye focuses the light. The lens in the eye can then focus the light on the retina.

Common Vision Problems

People who have normal vision can clearly see objects that are close and objects that are far away. They can also tell the difference between all colors of visible light. But because the eye is complex, it's no surprise that many people have defects in their eyes that affect their vision.

Nearsightedness and Farsightedness

The lens of a properly working eye focuses light on the retina. So, the images formed are always clear. Two common vision problems happen when light is not focused on the retina, as shown in **Figure 2. Nearsightedness** happens when a person's eye is too long. A nearsighted person can see something clearly only if it is nearby. Objects that are far away look blurry. **Farsightedness** happens when a person's eye is too short. A farsighted person can see faraway objects clearly. But things that are nearby look blurry. **Figure 2** also shows how these vision problems can be corrected with glasses.

nearsightedness a condition in which the lens of the eye focuses distant objects in front of rather than on the retina

farsightedness a condition in which the lens of the eye focuses distant objects behind rather than on the retina

✓ **Reading Check** What causes nearsightedness and farsightedness? (*See the Appendix for answers to Reading Checks.*)

BRAIN FOOD

Kepler Explains the Eye In 1604, Austrian mathematician and astronomer Johannes Kepler (1571–1630) gave the first correct explanation of how the human eye works, including why the image formed on the retina is upside down. Kepler's interest in light and optics led to his study of the eye.

Is That a Fact!

Some chickens wear red contact lenses. The lenses don't improve the chickens' vision—they just make the chickens see everything in red! Chickens that see in red are less aggressive and produce more eggs. But it is difficult to fit a chicken for contact lenses properly, and chickens often lose their contacts quickly.

Reteaching — **BASIC**

Diagramming the Eye Have students work in pairs to draw a diagram of the human eye. Students may use **Figure 1** in this section as a reference. When they finish their diagram, have the students take turns explaining the function of each part of the eye. **LS** Visual/Verbal

Quiz — **GENERAL**

1. Which vision problem occurs when the eye is too short? when it is too long? (farsightedness; nearsightedness)

2. In the eye, rods respond to movement and light but not to colors, while the cones detect colors. Which do you think is more important? Why? (Accept all reasonable answers. Sample answer: rods; It is more important to be able to see movement because you can see if something is coming toward you and know to get out of the way.)

Alternative Assessment — **GENERAL**

Concept Mapping Have students make a concept map by using the following terms: *retina, cornea, lens, optic nerve, farsightedness, nearsightedness, convex lens,* and *concave lens.* **LS** Verbal/Visual

Figure 3 *The photo on the left is what a person who has normal vision sees. The photo on the right is a simulation of what a person who has red-green color deficiency might see.*

Color Deficiency

About 5% to 8% of men and 0.5% of women in the world have *color deficiency,* or colorblindness. The majority of people who have color deficiency can't tell the difference between shades of red and green or can't tell red from green. **Figure 3** compares what a person with normal vision sees with what a person who has red-green color deficiency sees. Color deficiency cannot be corrected.

Color deficiency happens when the cones in the retina do not work properly. The three kinds of cones are named for the colors they detect most—red, green, or blue. But each kind can detect many colors of light. A person who has normal vision can see all colors of visible light. But in some people, the cones respond to the wrong colors. Those people see certain colors, such as red and green, as a different color, such as yellow.

✓ **Reading Check** What are the three kinds of cones?

CONNECTION TO Biology

Color Deficiency and Genes The ability to see color is a sex-linked genetic trait. Certain genes control which colors of light the cones detect. If these genes are defective in a person, that person will have color deficiency. A person needs one set of normal genes to have normal color vision. Genes that control the red cones and the green cones are on the X chromosome. Women have two X chromosomes, but men have only one. So, men are more likely than women to lack a set of these genes and to have red-green color deficiency. Research two other sex-linked traits, and make a graph comparing the percentage of men and women who have the traits. **ACTIVITY**

CONNECTION ACTIVITY Math — **GENERAL**

Color Deficiency Calculations Have students compare the number of males and females that have color deficiency. First, have students calculate how many men out of 600 have color deficiency. (600 × 5% = 30; 600 × 8% = 48; 30–48 men) Then, have them calculate how many women out of 600 have color deficiency. (600 × 0.5% = 3; 3 women) Finally, have students compare the numbers. (Ten to sixteen times more men than women have color deficiency.) **LS** Logical

Answer to Reading Check
The three kinds of cones are red, blue, and green.

Surgical Eye Correction

Using surgery to correct nearsightedness or farsightedness is possible. Surgical eye correction works by reshaping the patient's cornea. Remember that the cornea refracts light. So, reshaping the cornea changes how light is focused on the retina.

To prepare for eye surgery, an eye doctor uses a machine to measure the patient's corneas. A laser is then used to reshape each cornea so that the patient gains perfect or nearly perfect vision. **Figure 4** shows a patient undergoing eye surgery.

Risks of Surgical Eye Correction

Although vision-correction surgery can be helpful, it has some risks. Some patients report glares or double vision. Others have trouble seeing at night. Other patients lose vision permanently. People under 20 years old shouldn't have vision-correction surgery because their vision is still changing.

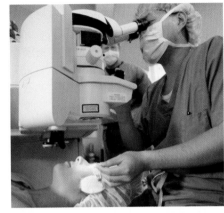

Figure 4 *An eye surgeon uses a very precise laser to reshape this patient's cornea.*

SECTION Review

Summary

- The human eye has several parts, including the cornea, the pupil, the iris, the lens, and the retina.
- Nearsightedness and farsightedness happen when light is not focused on the retina. Both problems can be corrected with glasses or eye surgery.
- Color deficiency is a condition in which cones in the retina respond to the wrong colors.
- Eye surgery can correct some vision problems.

Using Key Terms

1. Use each of the following terms in a separate sentence: *nearsightedness* and *farsightedness*.

Understanding Key Ideas

2. A person who is nearsighted will have the most trouble reading
   a. a computer screen in front of him or her.
   b. a book in his or her hands.
   c. a street sign across the street.
   d. the title of a pamphlet on a nearby table.

3. List the parts of the eye, and describe what each part does.

4. What are three common vision problems?

5. How are nearsightedness and farsightedness corrected?

6. Describe surgical eye correction.

7. What do the rods and cones in the eye do?

Math Skills

8. About 0.5% of women have a color deficiency. How many women out of 200 have a color deficiency?

Critical Thinking

9. **Forming Hypotheses** Why do you think color deficiency cannot be corrected?

10. **Expressing Opinions** Would you have surgical eye correction? Explain your reasons.

SCiLINKS

NSTA

Developed and maintained by the National Science Teachers Association

For a variety of links related to this chapter, go to www.scilinks.org

Topic: The Eye
SciLinks code: HSM0560

CHAPTER RESOURCES

Chapter Resource File

- Section Quiz GENERAL
- Section Review GENERAL
- Vocabulary and Section Summary GENERAL

Focus

Overview

This section describes some optical instruments that use lenses and mirrors. It also explains the way lasers work and discusses some uses for lasers. Students also learn about fiber optics, polarized light, and communication technology.

 **Bellringer**

Ask students to describe lasers. Then, ask students to list at least four uses for lasers.

Motivate

Demonstration — GENERAL

Optical Instruments If you have a camera and a classroom laser, demonstrate how they work. Ask students what the two items have in common and how they are different. Display other optical instruments, especially a microscope and a telescope, where students can see them. Discuss each instrument with students. Ask them what they know about each instrument and if they can explain how each one works. **LS** Visual

---

READING WARM-UP

Objectives

- Describe three optical instruments.
- Explain what laser light is, and identify uses for lasers.
- Describe how optical fibers work.
- Explain polarized light.
- Explain how radio waves and microwaves are used in four types of communication technology.

Terms to Learn

laser
hologram

READING STRATEGY

Prediction Guide Before reading this section, write the title of each heading in this section. Next, under each heading, write what you think you will learn.

Light and Technology

What do cameras, telescopes, lasers, cellular telephones, and satellite televisions have in common?

They are all types of technology that use light or other electromagnetic waves. Read on to learn how these and other types of light technology are useful in your everyday life.

Optical Instruments

Optical instruments are devices that use mirrors and lenses to help people make observations. Some optical instruments help you see things that are very far away. Others help you see things that are very small. Some optical instruments record images. The optical instrument that you are probably most familiar with is the camera.

Cameras

Cameras are used to record images. **Figure 1** shows the parts of a 35 mm camera. A digital camera has a lens, a shutter, and an aperture (AP uhr chuhr) like a 35 mm camera has. But instead of using film, a digital camera uses light sensors to record images. The sensors send an electrical signal to a computer in the camera. This signal contains data about the image that is stored in the computer, on a memory stick, card, or disk.

Figure 1   How a Camera Works

The **shutter** opens and closes behind the lens to control how much light enters the camera. The longer the shutter is open, the more light enters the camera.

The **lens** of a camera is a convex lens that focuses light on the film. Moving the lens focuses light from objects at different distances.

The **film** is coated with chemicals that react when they are exposed to light. The result is an image stored on the film.

The **aperture** is an opening that lets light into the camera. The larger the aperture is, the more light enters the camera.

CHAPTER RESOURCES

Chapter Resource File

- **Lesson Plan**
- **Directed Reading A** BASIC
- **Directed Reading B** SPECIAL NEEDS

Technology

Transparencies
- Bellringer
- How a Camera Works
- How Refracting and Reflecting Telescopes Work
- *LINK TO LIFE SCIENCE* Compound Light Microscope

CONNECTION ACTIVITY
Math ————— ADVANCED

Camera Numbers Two numbers are very important to photographers: shutter speed and f-stop. Have students research both numbers, and do a presentation to explain their findings to the class. Encourage students to be creative. **LS** Visual/Logical

Figure 2 **How Refracting and Reflecting Telescopes Work**

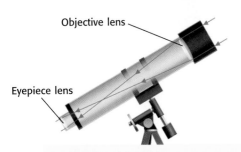

Objective lens

Eyepiece lens

A **refracting telescope** has two convex lenses. Light enters through the objective lens and forms a real image. This real image is then magnified by the eyepiece lens. You see this magnified image when you look through the eyepiece lens.

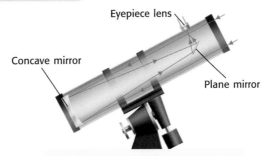

Eyepiece lens

Concave mirror

Plane mirror

A **reflecting telescope** has a concave mirror that collects and focuses light to form a real image. The light strikes a plane mirror that directs the light to the convex eyepiece lens, which magnifies the real image.

Telescopes

Telescopes are used to see detailed images of large, distant objects. Astronomers use telescopes to study things in space, such as the moon, planets, and stars. Telescopes are classified as either refracting or reflecting. *Refracting telescopes* use lenses to collect light. *Reflecting telescopes* use mirrors to collect light. **Figure 2** shows how these two kinds of telescopes work.

Light Microscopes

Simple light microscopes are similar to refracting telescopes. These microscopes have two convex lenses. An objective lens is close to the object being studied. An eyepiece lens is the lens you look through. Microscopes are used to see magnified images of tiny, nearby objects.

Lasers and Laser Light

A **laser** is a device that produces intense light of only one color and wavelength. Laser light is different from nonlaser light in many ways. One important difference is that laser light is *coherent*. When light is coherent, light waves move together as they travel away from their source. The crests and troughs of coherent light waves are aligned. So, the individual waves behave as one wave.

Reading Check What does it mean for light to be coherent? *(See the Appendix for answers to Reading Checks.)*

MATH PRACTICE

Microscope Magnification

Some microscopes use more than one lens to magnify objects. The power of each lens indicates the amount of magnification the lens gives. For example, a 10× lens magnifies objects 10 times. To find the amount of magnification given by two or more lenses used together, multiply the powers of the lenses. What is the magnification given by a 5× lens used with a 20× lens?

laser a device that produces intense light of only one wavelength and color

Answer to Reading Check
When light is coherent, light waves move together as they travel away from their source. Individual waves behave as one wave.

Using the Figure — GENERAL

Laser Light Use **Figure 3** to explain the two special properties of laser light: it is light of a single wavelength and color, and it is coherent. Light is coherent when the crests and troughs of light waves are aligned. This causes the individual light waves to act as a single wave. Nonlaser light sources emit light waves of many different wavelengths and colors, whose crests and troughs are not aligned. **LS Visual**

ACTIVITY — BASIC

Laser Model Have students make a model of a laser by using construction paper for the tube and aluminum foil for the mirrors. Students can use fishing weights for neon atoms and red "confetti" from a hole punch for the photons. Have students explain their model. **LS Kinesthetic**

Homework — GENERAL

Uses for Lasers Have students research the many ways argon lasers, carbon dioxide lasers, helium-neon lasers, and other lasers are used for medical diagnosis and treatment. **LS Verbal**

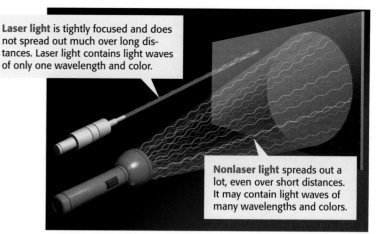

Laser light is tightly focused and does not spread out much over long distances. Laser light contains light waves of only one wavelength and color.

Figure 3 *Laser light is very different from nonlaser light.*

Nonlaser light spreads out a lot, even over short distances. It may contain light waves of many wavelengths and colors.

How Lasers Produce Light

Figure 3 compares laser and nonlaser light. The word *laser* stands for **l**ight **a**mplification by **s**timulated **e**mission of **r**adiation. *Amplification* is the increase in the brightness of the light. *Radiation* is energy transferred as electromagnetic waves.

What is stimulated emission? In an atom, an electron can move from one energy level to another. A photon (a particle of light) is released when an electron moves from a higher energy level to a lower energy level. The release of photons is called *emission*. *Stimulated emission* occurs when a photon strikes an atom that is in an excited state and makes the atom emit another photon. The newly emitted photon is identical to the first photon. The two photons travel away from the atom together. **Figure 4** shows how laser light is produced.

Figure 4   How a Helium-Neon Laser Works

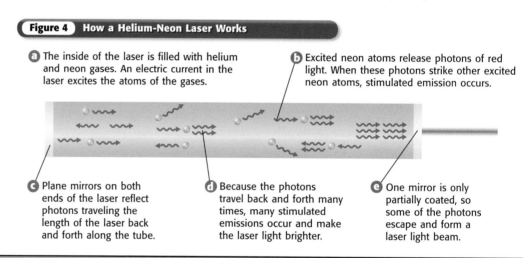

a The inside of the laser is filled with helium and neon gases. An electric current in the laser excites the atoms of the gases.

b Excited neon atoms release photons of red light. When these photons strike other excited neon atoms, stimulated emission occurs.

c Plane mirrors on both ends of the laser reflect photons traveling the length of the laser back and forth along the tube.

d Because the photons travel back and forth many times, many stimulated emissions occur and make the laser light brighter.

e One mirror is only partially coated, so some of the photons escape and form a laser light beam.

WEIRD SCIENCE

Laser light can be produced in a variety of ways. Gas lasers, such as the one in **Figure 4,** produce laser light from excited gas atoms. Solid-state lasers have a solid rather than a gas between the two mirrors. Semiconductor lasers use the same material that is found in computer chips to produce laser light. Most CD and DVD players use semiconductor lasers.

SCIENTISTS AT ODDS

The Laser Patent Dispute Two American physicists, Arthur Schawlow and Charles Townes, received a patent for the working principles of a laser in 1958. But Gordon Gould claimed that he not only had discovered how to produce laser light but also had named the process in 1957. Finally, in 1987, after many bitter court battles, Gould's claim was upheld.

Uses for Lasers

Lasers are used to make holograms, such as the one shown in **Figure 5**. A **hologram** is a piece of film that produces a three-dimensional image of an object. Holograms are similar to photographs because both are images recorded on film. However, unlike photographs, the images you see in holograms are not on the surface of the film. The images appear in front of or behind the film. If you move the hologram, you will see the image from different angles.

Lasers are also used for other tasks. For example, lasers are used to cut materials such as metal and cloth. Doctors sometimes use lasers for surgery. And CD players have lasers. Light from the laser in a CD player reflects off patterns on a CD's surface. The reflected light is converted to a sound wave.

✓ **Reading Check** How are holograms like photographs?

Optical Fibers

Imagine a glass thread that transmits more than 1,000 telephone conversations at the same time with flashes of light. This thread, called an *optical fiber,* is a thin, glass wire that transmits light over long distances. Some optical fibers are shown in **Figure 6**. Transmitting information through telephone cables is the most common use of optical fibers. Optical fibers are also used to network computers. And they allow doctors to see inside patients' bodies without performing major surgery.

Light in a Pipe

Optical fibers are like pipes that carry light. Light stays inside an optical fiber because of total internal reflection. *Total internal reflection* is the complete reflection of light along the inside surface of the material through which it travels. **Figure 6** shows total internal reflection in an optical fiber.

hologram a piece of film that produces a three-dimensional image of an object; made by using laser light

Figure 5 *Some holograms make three-dimensional images that look so real that you might want to reach out and touch them!*

Figure 6 | How Optical Fibers Work

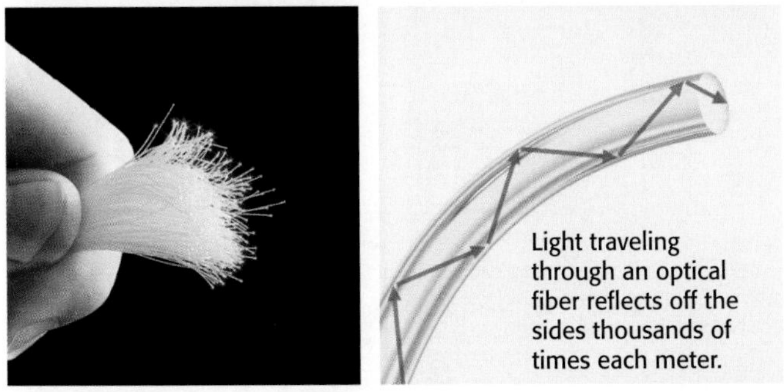

Light traveling through an optical fiber reflects off the sides thousands of times each meter.

Answer to Reading Check
Holograms are like photographs because both are images recorded on film.

CONNECTION to Life Science ——— GENERAL

Optical Tweezers Biologists use laser devices called optical tweezers to handle organisms without damaging them. Biologists also use optical tweezers to manipulate organelles within living cells without breaking the cell membrane, to move chromosomes within a cell nucleus, and to manipulate single strands of DNA.

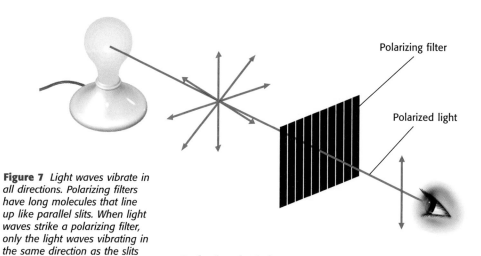

Demonstration — GENERAL

Polarizing Filter and Liquid Crystals Place a polarizing filter over the face of a liquid-crystal-display calculator. An overhead calculator gives excellent results. Rotate the filter; have students observe what happens. Ask students to speculate on what causes these changes. (As the filter is rotated, the display blacks out. This means there must be another polarizing filter inside the device because two polarizing filters are needed to block out light.) **LS** Visual

Liquid Crystals Polarized light is used in liquid crystal displays. When the polarized light passes through the liquid crystal material, the liquid crystals cause the polarized light to twist. When a weak electric voltage is applied to the liquid crystals, the polarized light no longer twists. This lets part of the display go dark and leaves a segment to form a number or letter.

Figure 7 *Light waves vibrate in all directions. Polarizing filters have long molecules that line up like parallel slits. When light waves strike a polarizing filter, only the light waves vibrating in the same direction as the slits pass through.*

Polarized Light

The next time you shop for sunglasses, look for some that have lenses that polarize light. Such sunglasses are good for reducing glare. *Polarized light* consists of light waves that vibrate in only one plane. **Figure 7** illustrates how light is polarized.

When light reflects off a horizontal surface, such as a car hood or a body of water, the light is polarized horizontally. You see this polarized light as glare. Polarizing sunglasses reduce glare from horizontal surfaces because the lenses have vertically polarized filters. These filters allow only vertically vibrating light waves to pass through them. Polarizing filters are also used by photographers to reduce glare in their photographs, as shown in **Figure 8.**

Figure 8 *These two photos were taken by the same camera and from the same angle. There is less reflected light in the photo at right because a polarizing filter was placed over the lens of the camera.*

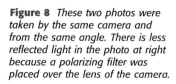

WEIRD SCIENCE

Three-dimensional (3-D) movies work by showing two images on the screen at the same time. The two images are similar but are filmed from slightly different angles. People watching the movie wear special glasses that allow each eye to see only one of the images. Because the eyes see slightly different images, the brain interprets the images as three-dimensional objects. The original 3-D glasses had one red lens and one blue lens that each filtered out one image. The 3-D glasses used today have polarized lenses. One lens allows only vertically polarized light to reach the eye, and the other lens allows only horizontally polarized light to reach the eye.

Blackout!

1. Hold a **lens from a pair of polarizing sunglasses** up to your eye, and look through the lens. Record your observations.

2. Put a **second polarizing lens** over the first lens. Make sure both lenses are right side up. Look through both lenses, and describe your observations.

3. Rotate one lens slowly as you look through both lenses, and describe what happens.

4. Why can't you see through the lenses when they are aligned a certain way?

Communication Technology

You may think that talking on the telephone has nothing to do with light. But if you are talking on a cordless telephone or a cellular telephone, you are using a form of light technology! Light is an electromagnetic wave. There are many different kinds of electromagnetic waves. Radio waves and microwaves are kinds of electromagnetic waves. And cordless telephones and cellular telephones use radio waves and microwaves to send signals.

Cordless Telephones

Cordless telephones are a combination of a regular telephone and a radio. There are two parts to a cordless telephone—the base and the handset. The base is connected to a telephone jack in the wall of a building. The base receives calls through the phone line. The base then changes the signal to a radio wave and sends the signal to the handset. The handset changes the radio signal to sound for you to hear. The handset also changes your voice to a radio wave that is sent back to the base.

✓ **Reading Check** What kind of electromagnetic wave does a cordless telephone use?

Cellular Telephones

The telephone in **Figure 9** is a cellular telephone. Cellular telephones are similar to the handset part of a cordless telephone because they send and receive signals. But a cellular telephone receives signals from tower antennas located across the country instead of from a base. And instead of using radio waves, cellular telephones use microwaves to send information.

Figure 9 *You can make and receive calls with a cellular telephone almost everywhere you go.*

Answer to Reading Check

A cordless telephone sends signals by using radio waves.

Close

Reteaching   **BASIC**

Polarized Bodies Line up desks into two rows that are 3 ft apart. At the end of the rows, have two students hold two broomsticks horizontally about 3 ft apart. Have students walk between the desks. Tell students that they are vertically polarized. Ask students what would happen if they tried to pass through the horizontally polarizing filter (the broomsticks). (They can't pass through the filter.) Then, ask students how a person must be positioned to travel through the horizontally polarizing filter. (He or she has to be lying down.) **LS Kinesthetic**

Quiz   **GENERAL**

1. Name three optical instruments. (microscopes, telescopes, cameras)

2. What is GPS? (*GPS* stands for *Global Positioning System*. It is a network of satellites that allow people to measure their position on Earth.)

3. How do polarizing lenses on sunglasses work? (They have filters that allow only vertically-polarized light waves to pass through, reducing the amount of glare.)

Alternative Assessment   **GENERAL**

Question Writing Have each student write seven questions with answers from this section. Play a game with the class. Each student should answer at least one question. **LS Verbal**

Satellite Television

Another technology that uses electromagnetic waves to transmit data is satellite television. Satellite television companies broadcast microwave signals from human-made satellites in space. Broadcasting from space allows more people to receive the signals than broadcasting from an antenna on Earth. Small satellite dishes on the roofs of houses or outside apartments collect the signals. The signals are then sent to the customer's television set. People who have satellite television usually have better TV reception than people who receive broadcasts from antennas on Earth.

The Global Positioning System

The Global Positioning System (GPS) is a network of 27 satellites that orbit Earth. These satellites continuously send microwave signals. The signals can be picked up by a GPS receiver on Earth and used to measure positions on the Earth's surface. **Figure 10** explains how GPS works. GPS was originally used by the United States military. But now, anyone in the world who has a GPS receiver can use the system. People use GPS to avoid getting lost and to have fun. Some cars have GPS road maps that can tell the car's driver how to get to a certain place. Hikers and campers use GPS receivers to find their way in the wilderness. And some people use GPS receivers for treasure-hunt games.

✓ *Reading Check* What are two uses for GPS?

CONNECTION TO Social Studies

Navigation GPS is a complex navigation system. Before GPS was developed, travelers and explorers used other techniques, such as compasses and stars, to find their way. Research an older form of navigation, and make a poster that summarizes what you learn.

ACTIVITY

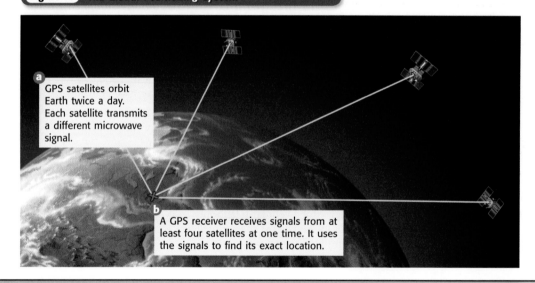

Figure 10 The Global Positioning System

a GPS satellites orbit Earth twice a day. Each satellite transmits a different microwave signal.

b A GPS receiver receives signals from at least four satellites at one time. It uses the signals to find its exact location.

Answer to Reading Check
Sample answer: GPS can be used by hikers and campers to find their way in the wilderness. GPS can also be used for treasure hunt games.

Is That a Fact!

Only 24 of the 27 GPS satellites are in operation at a given time. The other three are used as back-up in case one fails.

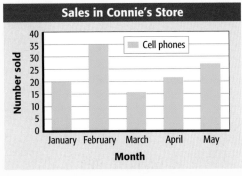

SECTION Review

Summary

- Optical instruments, such as cameras, telescopes, and microscopes, are devices that help people make observations.
- Lasers are devices that produce intense, coherent light of only one wavelength and color. Lasers produce light by a process called *stimulated emission*.
- Optical fibers transmit light over long distances.
- Polarized light contains light waves that vibrate in only one direction.

- Cordless telephones are a combination of a telephone and a radio. Information is transmitted in the form of radio waves between the handset and the base.
- Cellular phones transmit information in the form of microwaves to and from antennas.
- Satellite television is broadcast by microwaves from satellites in space.
- GPS is a navigation system that uses microwave signals sent by a network of satellites in space.

Using Key Terms

1. Use each of the following terms in a separate sentence: *laser* and *hologram*.

Understanding Key Ideas

2. Which of the following statements about laser light is NOT true?
   a. Laser light is coherent.
   b. Laser light contains light of only one wavelength.
   c. Laser light is produced by stimulated emission.
   d. Laser light spreads out over short distances.

3. List three optical instruments, and describe what they do.

4. What are four uses for lasers?

5. Describe how optical fibers work.

6. What is polarized light?

7. Describe two ways that satellites in space are useful in everyday life.

Critical Thinking

8. **Making Comparisons** Compare how a cordless telephone works with how a cellular telephone works.

9. **Making Inferences** Why do you think optical fibers can transmit information over long distances without losing much of the signal?

Interpreting Graphics

Use the graph below to answer the questions that follow.

Sales in Connie's Store

10. In which two months did Connie's store sell the most cellular telephones?

11. How many cellular telephones were sold in January?

SCILINKS®

NSTA
Developed and maintained by the National Science Teachers Association

For a variety of links related to this chapter, go to www.scilinks.org

Topic: Lasers
SciLinks code: HSM0853

CHAPTER RESOURCES

Chapter Resource File

- Section Quiz GENERAL
- Section Review GENERAL
- Vocabulary and Section Summary GENERAL
- Critical Thinking ADVANCED

Answers to Section Review

1. Sample answer: A laser can be used to perform surgeries. I have a hologram of a toy dragon.

2. d

3. A camera focuses light on a piece of film to record an image on the film. A telescope allows people to see objects that are far away. A microscope allows people to see enlarged images of objects that are very small.

4. Lasers are used to make holograms, to cut materials, to perform surgeries, and are used in CD players.

5. As light travels through an optical fiber, the light reflects off the sides of the fiber thousands of times each meter. The light stays in the fiber because of total internal reflection.

6. Polarized light is light that consists of waves that vibrate in only one plane.

7. Some satellites in space are used to broadcast television signals. Other satellites are part of the Global Positioning System and help people avoid getting lost.

8. Cordless telephones send and receive radio wave signals from a base that is connected to a telephone jack. Cellular telephones send and receive microwave signals from tower antennas located across the country.

9. Optical fibers can transmit information over long distances without losing much of the signal because of total internal reflection. Total internal reflection keeps the information from escaping from the fibers.

10. February and May

11. 20 cellular phones

Images from Convex Lenses

Teacher's Notes

Time Required

One or two 45-minute class periods

Lab Ratings

EASY ———————→ HARD

Teacher Prep △
Student Set-Up △
Concept Level △△△
Clean Up △

Safety Caution

Remind students to review all safety cautions and icons before beginning this lab activity.

Caution students about working near an open flame. Any loose hair or clothing should be tied back before beginning the experiment.

Lab Notes

Image 1 forms when the distance between the candle and the card is 4 times the focal length of the lens. However, students do NOT need to know the focal length of the lens to perform the procedure.

Images from Convex Lenses

A convex lens is thicker in the center than at the edges. Light rays passing through a convex lens come together at a focal point. Under certain conditions, a convex lens will create a real image of an object. This image will have certain characteristics, depending on the distance between the object and the lens. In this experiment, you will determine the characteristics of real images created by a convex lens—the kind of lens used as a magnifying lens.

Ask a Question

1 What are the characteristics of real images created by a convex lens? For example, are the images upright or inverted (upside down)? Are the images larger or smaller than the object?

Form a Hypothesis

2 Write a hypothesis that is a possible answer to the questions above. Explain your reasoning.

Test the Hypothesis

3 Copy the table below.

Data Collection				
Image	Orientation (upright/ inverted)	Size (larger/ smaller)	Image distance (cm)	Object distance (cm)
1				
2		DO NOT WRITE IN BOOK		
3				

OBJECTIVES

Use a convex lens to form images.

Determine the characteristics of real images formed by convex lenses.

MATERIALS

- candle
- card, index, 4 × 6 in. or larger
- clay, modeling
- convex lens
- jar lid
- matches
- meterstick

SAFETY

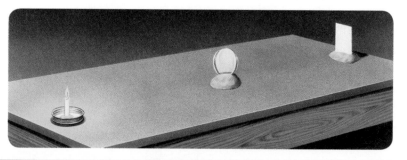

4. Use modeling clay to make a base for the lens. Place the lens and base in the middle of the table.

5. Stand the index card upright in some modeling clay on one side of the lens.

6. Place the candle in the jar lid, and anchor it with some modeling clay. Place the candle on the table so that the lens is halfway between the candle and the card. Light the candle.
   Caution: Use extreme care around an open flame.

7. In a darkened room, slowly move the card and the candle away from the lens while keeping the lens exactly halfway between the card and the candle. Continue until you see a clear image of the candle flame on the card. This is image 1.

8. Measure and record the distance between the lens and the card (image distance) and between the lens and the candle (object distance).

9. Is the image upright or inverted? Is it larger or smaller than the candle? Record this information in the table.

10. Move the lens toward the candle. The new object distance should be less than half the object distance measured in step 8. Move the card back and forth until you find a sharp image (image 2) of the candle on the card.

11. Repeat steps 8 and 9 for image 2.

12. Leave the card and candle in place and move the lens toward the card to get the third image (image 3).

13. Repeat steps 8 and 9 for image 3.

Analyze the Results

1. **Recognizing Patterns** Describe the trend between image distance and image size.

2. **Examining Data** What are the similarities between the real images that are formed by a convex lens?

Draw Conclusions

3. **Making Predictions** The lens of your eye is a convex lens. Use the information you collected to describe the image projected on the back of your eye when you look at an object.

Applying Your Data

Convex lenses are used in film projectors. Explain why your favorite movie stars are truly "larger than life" on the screen in terms of image distance and object distance.

Analyze the Results

1. When the image distance gets larger, the image size gets larger.

2. The images are inverted.

Draw Conclusions

3. The image projected on the back of your eye is a real image that is smaller than the object and inverted.

Applying Your Data

The object distance (from the film to the lens) is much smaller than the image distance (from the lens to the screen). Thus, the image projected on the screen is very large compared with the size of the image on the film itself.

Assignment Guide

SECTION	QUESTIONS
1	1, 4, 8, 11, 20
2	2–3, 6–7, 12, 15, 17
3	5, 9–10, 13–14, 18–19
1 and 3	16

ANSWERS

Using Key Terms

1. A concave mirror is a mirror shaped like the inside of a spoon.

2. Eye surgeons use a laser to reshape the cornea of an eye.

3. A person who has farsightedness has trouble reading a book.

4. A convex lens refracts light and focuses it inward to a focal point.

5. If you move a hologram around, you can see its three-dimensional image from different angles.

Understanding Key Ideas

6. c
7. b
8. c
9. c
10. a
11. c

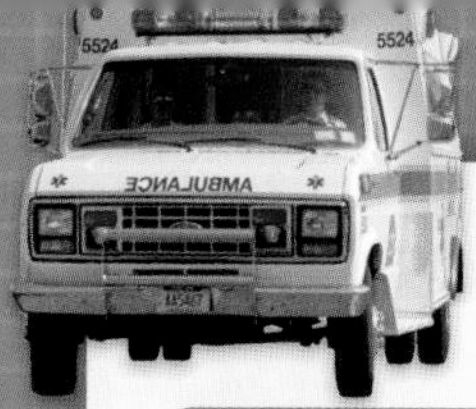

USING KEY TERMS

In each of the following sentences, replace the incorrect term with the correct term from the word bank.

nearsightedness   hologram
concave mirror   laser
plane mirror   convex lens
convex mirror   farsightedness

1 A convex mirror is a mirror shaped like the inside of a spoon.

2 Eye surgeons use a hologram to reshape the cornea of an eye.

3 A person who has nearsightedness has trouble reading a book.

4 A concave lens refracts light and focuses it inward to a focal point.

5 If you move a lens around, you can see its three-dimensional image from different angles.

UNDERSTANDING KEY IDEAS

Multiple Choice

6 Which of the following parts of the eye refracts light?

a. pupil   c. lens
b. iris   d. retina

7 A vision problem that happens when light is focused in front of the retina is

a. farsightedness.
b. nearsightedness.
c. color deficiency.
d. None of the above

8 What kind of mirror provides images of large areas and is used for security?

a. a plane mirror
b. a concave mirror
c. a convex mirror
d. All of the above

9 A simple refracting telescope has

a. a convex lens and a concave lens.
b. a concave mirror and a convex lens.
c. two convex lenses.
d. two concave lenses.

10 Light waves in a laser beam interact and act as one wave. This light is called

a. coherent light.   c. polarized light.
b. emitted light.   d. reflected light.

11 When you look at yourself in a plane mirror, you see a

a. real image behind the mirror.
b. real image on the surface of the mirror.
c. virtual image that appears to be behind the mirror.
d. virtual image that appears to be in front of the mirror.

Short Answer

12 What kind of eyeglass lens should be prescribed for a person who cannot focus on nearby objects? Explain.

13 How is a hologram different from a photograph?

14 Why might a scientist who is working at the North Pole need polarizing sunglasses?

Math Skills

15 Ms. Welch's class conducted a poll about vision problems. Of the 150 students asked, 21 reported that they are nearsighted. Six of the nearsighted students wear contact lenses to correct their vision, and the rest wear glasses.

a. What percentage of the students asked is nearsighted?

b. What percentage of the students asked wears glasses?

CRITICAL THINKING

16 **Concept Mapping** Use the following terms to create a concept map: *lens, telescope, camera, real image, virtual image,* and *optical instrument.*

17 **Analyzing Ideas** Stoplights are usually mounted so that the red light is on the top and the green light is on the bottom. Why is it important for a person who has red-green color deficiency to know this arrangement?

18 **Applying Concepts** How could you find out if a device that produces red light is a laser or if it is just a red flashlight?

19 **Making Inferences** Imagine that you have a GPS receiver. When you use your receiver in the park and are surrounded by tall trees, the receiver easily finds your location. But when you use your receiver downtown and are surrounded by tall buildings, the receiver cannot determine your location. Why do you think there is a difference in reception? Describe a situation in which poor GPS reception around tall buildings could cause problems.

INTERPRETING GRAPHICS

20 Look at the ray diagrams below. For each diagram, identify the type of mirror that is being used and the kind of image that is being formed.

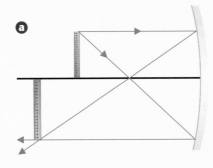

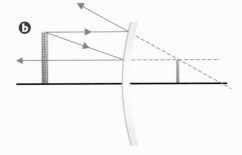

Standardized Test Preparation

Teacher's Note

To provide practice under more realistic testing conditions, give students 20 minutes to answer all of the questions in this Standardized Test Preparation.

MISCONCEPTION ALERT

Answers to the standardized test preparation can help you identify student misconceptions and misunderstandings.

READING

Passage 1

1. D
2. H
3. D

TEST DOCTOR

Question 2: Students may be tempted to select choice I because the passage states that Morgan patented the traffic signal and because the year 1920 is mentioned in the passage. However, the passage clearly states that Morgan patented the traffic signal in 1923. The correct choice is H.

READING

Read each of the passages below. Then, answer the questions that follow each passage.

Passage 1 One day in the 1920s, an automobile collided with a horse and carriage. Garrett Morgan witnessed this, and the accident gave him an idea. Morgan designed a signal that included signs to direct traffic at busy intersections. The signal could be seen from a distance and could be clearly understood. Morgan patented the first traffic signal in 1923. Unlike the small, three-bulb signal boxes used today, the early <u>versions</u> were T shaped and had the words *stop* and *go* printed on them.

Morgan's invention was an immediate success. Morgan sold the patent to General Electric Corporation for $40,000—a large sum in those days. Since then, later versions of Morgan's traffic signal have been a mainstay of traffic control.

1. In the passage, what does the word *versions* refer to?
   - **A** automobiles
   - **B** accidents
   - **C** light bulbs
   - **D** traffic signals

2. Which of the following statements is a fact?
   - **F** Morgan still makes money selling traffic signals today.
   - **G** Traffic signals were confusing and caused a lot of accidents.
   - **H** Morgan came up with the idea of a traffic signal after seeing a traffic accident.
   - **I** Morgan patented the traffic signal in 1920.

3. How were the first traffic signals similar to the signals used today?
   - **A** They were T shaped.
   - **B** They contained three light bulbs.
   - **C** The words *stop* and *go* were printed on them.
   - **D** They directed traffic at busy intersections.

Passage 2 Twenty years ago, stars were very visible, even above large cities. Now, the stars above large cities are <u>obscured</u> by the glow from city lights. This glow, called sky glow, is created when light reflects off dust and particles in the atmosphere. Sky glow is also called light pollution.

The majority of light pollution comes from outdoor lights, such as headlights, street lights, porch lights, and parking-lot lights. Unlike other kinds of pollution, light pollution can easily be reduced. For example, using covered outdoor lights keeps the light angled downward, which prevents most of the light from reaching particles in the sky. Also, using motion-sensitive lights and timed lights helps eliminate unnecessary light.

1. Which of the following **best** describes the reason the author wrote the passage?
   - **A** to explain light pollution and to explain how to reduce it
   - **B** to convince people to look at stars
   - **C** to explain why people should not live in cities
   - **D** to describe the beauty of sky glow

2. Which of the following contributes the least amount to light pollution?
   - **F** headlights on cars
   - **G** lights inside homes
   - **H** lights used in outdoor stadiums
   - **I** lights in large parking lots

3. In the passage, what does the word *obscured* mean?
   - **A** made brighter
   - **B** reflected
   - **C** polluted
   - **D** made difficult to see

Passage 2

1. A
2. G
3. D

TEST DOCTOR

Question 3: Students may find answering this question difficult because no definition for the word *obscured* is given. However, students should note that the first two sentences in the passage describe a change from 20 years ago to the present. According to the first sentence, stars were very visible 20 years ago. Therefore, the implied change is that the stars are less visible in the present. Thus, choice D is correct.

INTERPRETING GRAPHICS

The table below shows details about four lasers sold by a laser company. Use the table below to answer the questions that follow.

Laser Specifications

Color	Power (mW)	Wavelength (nm)	Mass (kg)
Blue	15	488	2.8
Yellow	5	568	5.8
Red	18	633	0.9
Red	10	633	0.6

1. What is the mass of the laser that has the most power?

A 0.6 kg

B 0.9 kg

C 2.8 kg

D 5.8 kg

2. The company also sells a laser that has a wavelength of 633 nm and a power of 5 mW. Which of the following statements **best** predicts the mass of this laser?

F The laser has a mass of less than 0.6 kg.

G The laser has a mass between 0.6 kg and 0.9 kg.

H The laser has a mass greater than 0.9 kg.

I The laser has a mass of 5.8 kg.

3. Based on the information in the table, which statement is most likely true?

A The power of the laser determines the color of light.

B The wavelength of the laser determines the color of light.

C The mass of the laser determines the color of light.

D There is not enough information to determine the answer.

MATH

Read each question below, and choose the best answer.

1. Micah has a box that has a length of 16 cm, a width of 10 cm, and a height of 5 cm. What is the volume of the box?

A 1,600 cm³

B 800 cm³

C 700 cm³

D 500 cm³

2. The table below shows the low temperature in Minneapolis, Minnesota, for five days in December.

Day	Temperature (°C)
Monday	−12
Tuesday	−8
Wednesday	7
Thursday	−3
Friday	11

Which list shows the temperatures from lowest to highest?

F −3°C, −8°C, −12°C, 7°C, 11°C

G −3°C, 7°C, −8°C, 11°C, −12°C

H −12°C, 11°C, 7°C, −8°C, −3°C

I −12°C, −8°C, −3°C, 7°C, 11°C

3. The power of a microscope lens is the amount of magnification the lens gives. For example, a 10× lens magnifies objects 10 times. How many times is an object magnified if it is viewed with both a 5× lens and a 30× lens?

A 35 times

B 60 times

C 150 times

D 350 times

Standardized Test Preparation

INTERPRETING GRAPHICS

1. B

2. F

3. B

 TEST DOCTOR

Question 3: To answer this question, students must focus their attention on the last two rows of the table because the colors of the lasers in the last two rows are the same. When students look across these two rows, it should be apparent that the only other factor that does not change is the wavelength of the lasers. Therefore, B is the correct choice.

MATH

1. B

2. I

3. C

TEST DOCTOR

Question 3: To answer this question, students can use the fact that the total magnification given by two or more lenses used at the same time is equal to the product of the powers of the lenses. (That is, students must multiply the powers of the lenses together.) The correct answer is C: $(5\times) \times (30\times) = 150\times$.

Students may also deduce the answer to this question by noting that the second lens magnifies the image from the first lens and that the image from the first lens is already magnified. So, the object is first magnified 5 times and then is magnified 30 more times for a total of 150 times.

CHAPTER RESOURCES

Chapter Resource File

• Standardized Test Preparation **GENERAL**

State Resources

For specific resources for your state, visit **go.hrw.com** and type in the keyword **HSMSTR**.

Science, Technology, and Society

Background

At this point, bionic eyes are still in the testing and research stage. Scientists have already tested silicon-based bionic eyes on several people. One patient was even able to drive a car on a test course. However, the early version of the silicon chip has only 16 electrodes, so it does not deliver as clear an image as a healthy eye, which has millions of cells. At this level, patients could expect to be able to distinguish between light and dark areas and to be able to see general shapes. In the future, scientists expect to be able to use chips that have 1,000 electrodes or more.

Scientific Debate

Discussion ——— GENERAL

Encourage students to express whether they believe cell phones are harmful. Ask students to explain why they came to that conclusion.

Science in Action

Science, Technology, and Society

Bionic Eyes

Imagine bionic eyes that allow a person who is blind to see. Researchers working on artificial vision think that the technology will be available soon. Many companies are working on different ways to restore sight to people who are blind. Some companies are developing artificial corneas, while other companies are building artificial retinas. One item that has already been tested on people is a pair of glasses that provides limited vision. The glasses have a camera that sends a signal to an electrode implanted in the person's brain. The images are black and white and are not detailed, but the person who is wearing the glasses can see obstacles in his or her path.

Language Arts ACTIVITY

WRITING SKILL Write a one-page story about a teen who has his or her eyesight restored by a bionic eye. What would the teen want to see first? What would the teen do that he or she couldn't do before?

Scientific Debate

Do Cellular Telephones Cause Cancer?

As cellular telephones became popular, people began to wonder if the phones were dangerous. Some cell-phone users claimed that the microwave energy from their cell phones caused them to develop brain cancer. So far, most research shows that the microwave energy emitted by cell phones is too low and too weak to damage human tissue. However, some studies have shown negative effects. There is some evidence that the low-power microwave energy used by cell phones may damage DNA and may cause cells to shrink. Because so many people use cellular phones, research continues around the world.

Math ACTIVITY

The American Cancer Society estimates that 0.006% of people in the United States will be diagnosed with brain cancer each year. If a city has a population of 50,000 people, how many people in that city will be diagnosed with brain cancer in one year?

Answer to Language Arts Activity

Accept all reasonable answers. All students' stories should describe what the teen wants to see first and what the teen wants to do once his or her sight is restored. For example, a teen may want to see his family first and then may want to try playing a sport like soccer.

Answer to Math Activity

Three people in that city will be diagnosed with brain cancer in 1 year (50,000 × 0.006% = 3).

People in Science

Sandra Faber

Astronomer What do you do when you send a telescope into space and then find out that it is broken? You call Dr. Sandra Faber, a professor of astronomy at the University of California, Santa Cruz (UCSC). In April 1990, after the *Hubble Space Telescope* went into orbit, scientists found that the images the telescope collected were not turning out as expected. Dr. Faber's team at UCSC was in charge of a device on *Hubble* called the *Wide Field Planetary Camera*. Dr. Faber and her team decided to test the telescope to determine what was wrong.

To perform the test, they centered *Hubble* onto a bright star and took several photos. From those photos, Dr. Faber's team created a model of what was wrong. After reporting the error to NASA and presenting the model they had developed, Dr. Faber and a group of experts began to correct the problem. The group's efforts were a success and put *Hubble* back into operation so that astronomers could continue researching stars and other objects in space.

Social Studies ACTIVITY

Research the history of the telescope. Make a timeline with the dates of major events in telescope history. For example, you could include the first use of a telescope to see the rings of Saturn in your timeline.

go.hrw.com

To learn more about these Science in Action topics, visit go.hrw.com and type in the keyword **HP5LOWF**.

Current Science

Check out Current Science® articles related to this chapter by visiting go.hrw.com. Just type in the keyword **HP5CS23**.

Background

In addition to working on the *Hubble Space Telescope*, Dr. Faber has played a key role in the development of the Keck telescopes in Hawaii, and she is currently working on *DEIMOS*, the *Deep Imaging Multi-Object Spectrograph*. A spectrograph is a device used to spread light gathered by the telescope into its different wavelengths, or colors. Depending on which wavelengths are brighter, you can determine what elements are present in the distant object you are looking at. You not only can tell what galaxies are made of but also can determine red shift. Knowing the red shift of galaxies allows you to know how far away they are, how fast they are moving, and what their relative masses are.

Dr. Faber compares the astronomer's method of studying the universe to a geologist using a core sample to study layers of sediment. "When you take a picture with the *Hubble*—the deepest picture of the universe ever taken—it's like a pencil-sized beam shooting out into space," she says. "In the foreground you see these recent things, and in the background you see the very early things. It's kind of a core drilling in space—but also a core drilling in time."

Answer to Social Studies Activity

Accept all reasonable responses. All students should create a timeline of major events in telescope history. Students may include events such as the invention of the telescope, the use of telescopes to identify the planet Pluto, and the launch of the *Hubble Space Telescope*.

Wave Speed, Frequency, and Wavelength

Teacher's Notes

Time Required

One or two 45-minute class periods

Lab Ratings

EASY			HARD

Teacher Prep 🧪
Student Set-Up 🧪
Concept Level 🧪🧪
Clean Up 🧪

MATERIALS

The materials listed for this lab are for each group of 3 students. This lab can also be done as a teacher demonstration if space or materials are limited.

Safety Caution

Remind students to review all safety cautions and icons before beginning this lab activity.

Lab Notes

You may need to demonstrate step 3 in Part A. Do all of Part A before beginning Part B.

Skills Practice Lab

Wave Speed, Frequency, and Wavelength

Wave speed, frequency, and wavelength are three related properties of waves. In this lab, you will make observations and collect data to determine the relationship among these properties.

MATERIALS

- meterstick
- stopwatch
- toy, coiled spring

SAFETY

Part A: Wave Speed

Procedure

1. Copy Table 1.

Table 1 Wave Speed Data			
Trial	**Length of spring (m)**	**Time for wave (s)**	**Speed of wave (m/s)**
1			
2			
3		DO NOT WRITE IN BOOK	
Average			

2. Two students should stretch the spring to a length of 2 m to 4 m on the floor or on a table. A third student should measure the length of the spring. Record the length in Table 1.

3. One student should pull part of the spring sideways with one hand, as shown at right, and release the pulled-back portion. This action will cause a wave to travel down the spring.

4. Using a stopwatch, the third student should measure how long it takes for the wave to travel down the length of the spring and back. Record this time in Table 1.

5. Repeat steps 3 and 4 two more times.

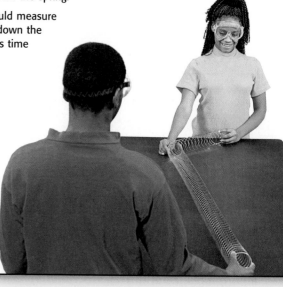

CHAPTER RESOURCES

Chapter Resource File

- Datasheet for LabBook
- Lab Notes and Answers

Part B: Wavelength and Frequency

Procedure

1. Keep the spring the same length that you used in Part A.

2. Copy Table 2.

Table 2 Wavelength and Frequency Data				
Trial	Length of spring (m)	Time for 10 cycles (s)	Wave frequency (Hz)	Wavelength (m)
1				
2				
3		DO NOT WRITE IN BOOK		
Average				

3. One of the two students holding the spring should start shaking the spring from side to side until a wave pattern appears that resembles one of those shown.

4. Using the stopwatch, the third student should measure and record how long it takes for 10 cycles of the wave pattern to occur. (One back-and-forth shake is 1 cycle.) Keep the pattern going so that measurements for three trials can be made.

Analyze the Results

Part A

1. Calculate and record the wave speed for each trial. (Speed equals distance divided by time; distance is twice the spring length.)

2. Calculate and record the average time and the average wave speed.

Part B

3. Calculate the frequency for each trial by dividing the number of cycles (10) by the time. Record the answers in Table 2.

4. Determine the wavelength using the equation at right that matches your wave pattern. Record your answer in Table 2.

5. Calculate and record the average time and frequency.

Draw Conclusions: Parts A and B

6. Analyze the relationship among speed, wavelength, and frequency. Multiply or divide any two of them to see if the result equals the third. (Use the averages from your data tables.) Write the equation that shows the relationship.

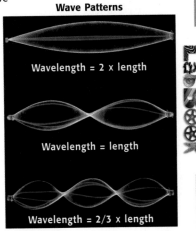

Wave Patterns

Wavelength = 2 x length

Wavelength = length

Wavelength = 2/3 x length

The Speed of Sound

Teacher's Notes

Time Required

One 45-minute class period

Lab Ratings

EASY ——————————→ HARD

Teacher Prep 🧪🧪
Student Set-Up 🧪🧪
Concept Level 🧪🧪
Clean Up 🧪

MATERIALS

Students will need stopwatches, measuring tapes, and various types of noisemakers (cymbals, wood blocks, drums, horns, and so on). You will need an area where students can hear an echo. A long hall will do, but this lab generally works better outside. To make this lab even more interesting, you might have a contest with awards for the most creative experiment, the most accurate experiment, and the experiment with the least spread in the data for multiple measurements.

Safety Caution

Be sure that you have approved all experimental designs before students proceed.

Inquiry Lab

The Speed of Sound

In the chapter entitled "The Nature of Sound," you learned that the speed of sound in air is 343 m/s at 20°C (approximately room temperature). In this lab, you'll design an experiment to measure the speed of sound yourself—and you'll determine if you're "up to speed"!

MATERIALS

• items to be determined by the students and approved by the teacher

Procedure

1. Brainstorm with your teammates to come up with a way to measure the speed of sound. Consider the following as you design your experiment:

   a. You must have a method of making a sound. Some simple examples include speaking, clapping your hands, and hitting two boards together.

   b. Remember that speed is equal to distance divided by time. You must devise methods to measure the distance that a sound travels and to measure the amount of time it takes for that sound to travel that distance.

   c. Sound travels very rapidly. A sound from across the room will reach your ears almost before you can start recording the time! You may wish to have the sound travel a long distance.

   d. Remember that sound travels in waves. Think about the interactions of sound waves. You might be able to include these interactions in your design.

2. Discuss your experimental design with your teacher, including any equipment you need. Your teacher may have questions that will help you improve your design.

3. Once your design is approved, carry out your experiment. Be sure to perform several trials. Record your results.

Analyze the Results

1. Was your result close to the value given in the introduction to this lab? If not, what factors may have caused you to get such a different value?

2. Why was it important for you to perform several trials in your experiment?

Draw Conclusions

3. Compare your results with those of your classmates. Determine which experimental design provided the best results. Explain why you think this design was so successful.

CHAPTER RESOURCES

Chapter Resource File

• Datasheet for LabBook
• Lab Notes and Answers

Paul Boyle
Perry Heights Middle School
Evansville, Indiana

Analyze the Results

1. Answers may vary. Factors that could cause a different value include air temperature and accuracy of distance and time measurements.

2. Several trials are necessary in order to confirm results and to rule out the possibility of reporting an error because just one measurement was used.

Draw Conclusions

3. Answers may vary. Accept all reasonable responses.

Skills Practice Lab

Tuneful Tube

If you have seen a singer shatter a crystal glass simply by singing a note, you have seen an example of resonance. For the glass to shatter, the note has to match the resonant frequency of the glass. A column of air within a cylinder can also resonate if the air column is the proper length for the frequency of the note. In this lab, you will investigate the relationship between the length of an air column, the frequency, and the wavelength during resonance.

MATERIALS

- eraser, pink, rubber
- graduated cylinder, 100 mL
- paper, graph
- plastic tube, supplied by your teacher
- ruler, metric
- tuning forks, different frequencies (4)
- water

SAFETY

Procedure

1 Copy the data table below.

Data Collection Table			
Frequency (Hz)			
Length (cm)	DO NOT WRITE IN BOOK		

2 Fill the graduated cylinder with water.

3 Hold a plastic tube in the water so that about 3 cm is above the water.

4 Record the frequency of the first tuning fork. Gently strike the tuning fork with the eraser, and hold the tuning fork so that the prongs are just above the tube, as shown at right. Slowly move the tube and fork up and down until you hear the loudest sound.

5 Measure the distance from the top of the tube to the water. Record this length in your data table.

6 Repeat steps 3–5 using the other three tuning forks.

Analyze the Results

1 Calculate the wavelength (in centimeters) of each sound wave by dividing the speed of sound in air (343 m/s at 20°C) by the frequency and multiplying by 100.

2 Make the following graphs: air column length versus frequency and wavelength versus frequency. On both graphs, plot the frequency on the *x*-axis.

3 Describe the trend between the length of the air column and the frequency of the tuning fork.

4 How are the pitches you heard related to the wavelengths of the sounds?

CHAPTER RESOURCES

Chapter Resource File

- **Datasheet for LabBook**
- **Lab Notes and Answers**

Jennifer Ford
North Ridge Middle School
North Richland Hills, Texas

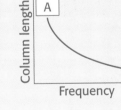

LabBook

Skills Practice Lab

Tuneful Tube

Teacher's Notes

Time Required

One 45-minute class period

Lab Ratings

EASY	→		HARD

Teacher Prep 🧪🧪🧪
Student Set-Up 🧪
Concept Level 🧪🧪🧪
Clean Up 🧪🧪

MATERIALS

The length of tube needed (PVC pipe works well) equals 25 times the speed of sound in air (343 m/s at 20°C) divided by the lowest frequency of the tuning forks used. (The answer is one-fourth the wavelength of the sound wave expressed in centimeters.) Cut each tube at least 5 cm longer than the calculated length. You may want to demonstrate the resonance point so students can hear the change in volume.

Analyze the Results

1. Answers will depend on the frequency of the tuning forks.

2. Graphs will depend on results. Sample graphs are shown at left.

3. As the frequency decreases, the length of the air column increases.

4. The pitches (which are determined by the frequencies) are inversely related to the wavelengths of the sounds. As the pitch gets lower (when the frequency decreases), the wavelength increases.

The Energy of Sound

Teacher's Notes

Time Required

One or two 45-minute class periods

Lab Ratings

🧪   🧪🧪   🧪🧪🧪   🧪🧪🧪🧪

EASY ———————→ HARD

Teacher Prep 🧪🧪
Student Set-Up 🧪
Concept Level 🧪🧪🧪🧪
Clean Up 🧪

Safety Caution

Make sure students are at a safe distance before you perform the Part D exercise. You might wish to take the class outside.

Skills Practice Lab

The Energy of Sound

In the chapter entitled "The Nature of Sound," you learned about various properties and interactions of sound. In this lab, you will perform several activities that will demonstrate that the properties and interactions of sound all depend on one thing—the energy carried by sound waves.

Part A: Sound Vibrations

Procedure

1 Lightly strike a tuning fork with the eraser. Slowly place the prongs of the tuning fork in the plastic cup of water. Record your observations.

Part B: Resonance

Procedure

1 Strike a tuning fork with the eraser. Quickly pick up a second tuning fork in your other hand, and hold it about 30 cm from the first tuning fork.

2 Place the first tuning fork against your leg to stop the tuning fork's vibration. Listen closely to the second tuning fork. Record your observations, including the frequencies of the two tuning forks.

3 Repeat steps 1 and 2, using the remaining tuning fork as the second tuning fork.

Part C: Interference

Procedure

1 Use the two tuning forks that have the same frequency, and place a rubber band tightly over the prongs near the base of one tuning fork, as shown at right. Strike both tuning forks against the eraser. Hold the stems of the tuning forks against a table, 3 cm to 5 cm apart. If you cannot hear any differences, move the rubber band up or down the prongs. Strike again. Record your observations.

MATERIALS

- cup, plastic, small, filled with water
- eraser, pink, rubber
- rubber band
- string, 50 cm
- tuning forks, same frequency (2), different frequency (1)

SAFETY

CHAPTER RESOURCES

Chapter Resource File

- Datasheet for LabBook
- Lab Notes and Answers

Part D: The Doppler Effect

Procedure

1. Your teacher will tie the piece of string securely to the base of one tuning fork. Your teacher will then strike the tuning fork and carefully swing the tuning fork in a circle overhead. Record your observations.

Analyze the Results

1. How do your observations demonstrate that sound waves are carried through vibrations?

2. Explain why you can hear a sound from the second tuning fork when the frequencies of the tuning forks used are the same.

3. When using tuning forks of different frequencies, would you expect to hear a sound from the second tuning fork if you strike the first tuning fork harder? Explain your reasoning.

4. Did you notice the sound changing back and forth between loud and soft? A steady pattern like this one is called a *beat frequency.* Explain this changing pattern of loudness and softness in terms of interference (both constructive and destructive).

5. Did the tuning fork make a different sound when your teacher was swinging it than when he or she was holding it? If yes, explain why.

6. Is the actual pitch of the tuning fork changing when it is swinging? Explain.

Draw Conclusions

7. Explain how your observations from each part of this lab verify that sound waves carry energy from one point to another through a vibrating medium.

8. Particularly loud thunder can cause the windows of your room to rattle. How is this evidence that sound waves carry energy?

4. The loudness corresponds to constructive interference (when the crests of the sound waves overlap, increasing the amplitude), and the softness corresponds to destructive interference (when the crests and troughs of sound waves overlap, decreasing the amplitude). A beat frequency caused by constructive and destructive interference could be heard in Part C.

5. yes; In Part D, as the tuning fork swings toward the listeners, the pitch is higher because the sound waves in front of it are closer together and therefore have a higher frequency. As the tuning fork swings away from the listeners, the pitch is lower because the sound waves are farther apart and therefore have a lower frequency.

6. yes; The pitch you hear in Part D changes because of the Doppler effect, but the actual frequency of the tuning fork does not change.

Draw Conclusions

7. Part A shows that the vibrations of the tuning fork have energy that does work on the water. Part B shows that the energy from one vibrating tuning fork can be passed by vibrations through the air to cause another tuning fork to vibrate. Part C shows that the energy from each vibrating tuning fork can travel through the air as waves that can interfere with each other. Part D shows that the vibrations from a tuning fork travel through the air to your ears, and the amount of energy being carried by the vibration determines what is heard (higher pitch = higher frequency = higher energy).

8. It takes energy to move the windows to cause them to rattle. Therefore, energy from the thunder's sound waves must be transferred through the air to the windows.

Analyze the Results

1. In Part A, the water begins to move when the tuning fork is placed in the water. Vibrations from the tuning fork caused the water's movement.

2. The vibrating tuning fork causes the air to vibrate at a certain frequency. The energy of the vibrations is transferred through the air to the second tuning fork, which starts to resonate (vibrate at the same frequency). This phenomenon happened in Part B.

3. Hitting the first tuning fork harder causes a larger amount of energy to be transferred from the tuning fork to the air. However, the vibration of the air particles is not at the same frequency as the second tuning fork and will therefore not cause the second tuning fork to make a sound.

Skills Practice Lab

What Color of Light Is Best for Green Plants?

Teacher's Notes

Time Required

One 45-minute class period, plus 5 minutes/day for 5 days

Lab Ratings

EASY ———————————→ HARD

Teacher Prep 🧪🧪🧪
Student Set-Up 🧪🧪
Concept Level 🧪🧪🧪🧪
Clean Up 🧪

MATERIALS

Seeds may be purchased at a garden store and should be germinated several days before this activity. Use at least green and red lights. Colored bulbs or bulbs with theatrical gels can be used. For best results, give seedlings light of only one color.

Safety Caution

Caution students to use care when working near hot bulbs.

Analyze the Results

1. Other than natural light, red light is the best, and green light is the worst.

Skills Practice Lab

What Color of Light Is Best for Green Plants?

Plants grow well outdoors under natural sunlight. However, some plants are grown indoors under artificial light. A variety of colored lights are available for helping plants grow indoors. In this experiment, you'll test several colors of light to discover which color best meets the energy needs of green plants.

MATERIALS

- bean seedlings
- colored lights, supplied by your teacher
- marker, felt-tip
- paper towels
- Petri dishes with covers
- tape, masking
- water

SAFETY

Ask a Question

1. Which color of light is best for growing green plants?

Form a Hypothesis

2. Write a hypothesis that answers the question above. Explain your reasoning.

Test the Hypothesis

3. Use the masking tape and marker to label the side of each Petri dish with your name and the type of light under which you will place the dish.

4. Place a moist paper towel in each Petri dish. Place 5 seedlings on top of the paper towel. Cover each dish.

5. Record your observations of the seedlings, such as length, color, and number of leaves.

6. Place each dish under the appropriate light.

7. Observe the Petri dishes every day for at least 5 days. Record your observations.

Analyze the Results

1. Based on your results, which color of light is best for growing green plants? Which color of light is worst?

Draw Conclusions

2. Remember that the color of an opaque object (such as a plant) is determined by the colors the object reflects. Use this information to explain your answer to question 1 above.

3. Would a purple light be good for growing purple plants? Explain.

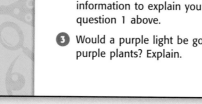

Draw Conclusions

2. Green plants reflect green light and use other colors to grow. Red is best because it is reflected the least.

3. Purple plants do not grow well under purple light because all the light is reflected.

CHAPTER RESOURCES

Chapter Resource File

- Datasheet for LabBook
- Lab Notes and Answers

Edith C. McAlanis
Socorro Middle School
El Paso, Texas

Skills Practice Lab

Which Color Is Hottest?

Will a navy blue hat or a white hat keep your head warmer in cool weather? Colored objects absorb energy, which can make the objects warmer. How much energy is absorbed depends on the object's color. In this experiment, you will test several colors under a bright light to determine which colors absorb the most energy.

Procedure

1 Copy the table below. Be sure to have one column for each color of paper you use and enough rows to end at 3 min.

Data Collection Table				
Time (s)	White	Red	Blue	Black
0				
15				
30				
45				
etc.				

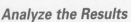 DO NOT WRITE IN BOOK

2 Tape a piece of colored paper around the bottom of a thermometer, and hold it under the light source. Record the temperature every 15 s for 3 min.

3 Cool the thermometer by removing the piece of paper and placing the thermometer in the cup of room-temperature water. After 1 min, remove the thermometer, and dry it with a paper towel.

4 Repeat steps 2 and 3 with each color, making sure to hold the thermometer at the same distance from the light source.

Analyze the Results

1 Prepare a graph of temperature (*y*-axis) versus time (*x*-axis). Using a different colored pencil or pen for each set of data, plot all data on one graph.

2 Rank the colors you used in order from hottest to coolest.

Draw Conclusions

3 Compare the colors, based on the amount of energy each absorbs.

4 In this experiment, a white light was used. How would your results be different if you used a red light? Explain.

5 Use the relationship between color and energy absorbed to explain why different colors of clothing are used for different seasons.

Analyze the Results

1. A sample graph is shown at right. (Red and blue may be reversed.)

Temp (°C) vs Time (s): Black, Red, Blue, White

2. black, red, blue, and white

Draw Conclusions

3. Black absorbs the most energy. Red absorbs more energy than blue. White absorbs the least energy.

4. The red and white papers would be coolest because they reflect red light.

5. Sample answer: Black clothes absorb energy and can help keep a person warm in winter. White clothes reflect energy and can help keep a person cool in summer.

Mirror Images

Teacher's Notes

Time Required

One or two 45-minute class periods

Lab Ratings

EASY ———————————————→ HARD

Teacher Prep 🧪
Student Set-Up 🧪
Concept Level 🧪🧪
Clean Up 🧪

MATERIALS

Mirrors with a focal length around 50 cm work well.

Safety Caution

Caution students about working near an open flame. Any loose hair or clothing should be tied back before beginning the experiment.

Skills Practice Lab

Mirror Images

When light actually passes through an image, the image is a **real image**. When light does not pass through the image, the image is a **virtual image**. Recall that plane mirrors produce only virtual images because the image appears to be behind the mirror where no light can pass through it.

In fact, all mirrors can form virtual images, but only some mirrors can form real images. In this experiment, you will explore the virtual images formed by concave and convex mirrors, and you will try to find a real image using both types of mirrors.

Part A: Finding Virtual Images

Procedure

1. Hold the convex mirror at arm's length away from your face. Observe the image of your face in the mirror.

2. Slowly move the mirror toward your face, and observe what happens to the image. Record your observations.

3. Move the mirror very close to your face. Record your observations.

4. Slowly move the mirror away from your face, and observe what happens to the image. Record your observations.

5. Repeat steps 1 through 4 with the concave mirror.

Analyze the Results

1. For each mirror, did you find a virtual image? How can you tell?

2. Describe the images you found. Were they smaller than, larger than, or the same size as your face? Were they right side up or upside down?

Draw Conclusions

3. Describe at least one use for each type of mirror. Be creative, and try to think of inventions that might use the properties of the two types of mirrors.

Part A

Analyze the Results

1. Students should find a virtual image for both mirrors. The virtual image appears to be behind the mirror, where no light rays can pass through the image.

2. Convex mirror images are smaller and upright; Concave mirror images are larger and upright when the mirror is close and smaller and upside down when the mirror is farther away.

Draw Conclusions

3. Accept all reasonable answers. Sample answer: Convex mirrors are used for wide-angle views in side-view mirrors on cars and to see around corners in busy hallways. Concave mirrors are used for makeup and shaving mirrors. Concave mirrors are also used in telescopes.

Part B: Finding a Real Image

Procedure

1 In a darkened room, place a candle in a jar lid near one end of a table. Use modeling clay to hold the candle in place. Light the candle. **Caution:** Use extreme care around an open flame.

2 Use more modeling clay to make a base to hold the convex mirror upright. Place the mirror at the other end of the table, facing the candle.

3 Hold the index card between the candle and the mirror but slightly to one side so that you do not block the candlelight, as shown below.

4 Move the card slowly from side to side and back and forth to see whether you can focus an image of the candle on it. Record your results.

5 Repeat steps 2–4 with the concave mirror.

Analyze the Results

1 For each mirror, did you find a real image? How can you tell?

2 Describe the real image you found. Was it smaller than, larger than, or the same size as the object? Was it right side up or upside down?

Draw Conclusions

3 Astronomical telescopes use large mirrors to reflect light to form a real image. Based on your results, do you think a concave or a convex mirror would be better for this instrument? Explain your answer.

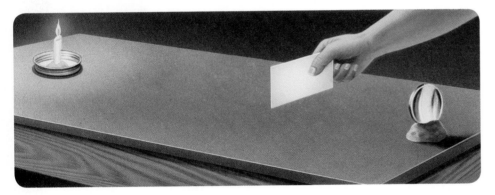

Kevin McCurdy
Elmwood Junior High
Rogers, Arkansas

CHAPTER RESOURCES

Chapter Resource File

- Datasheet for LabBook
- Lab Notes and Answers

Part B

Analyze the Results

1. A real image can be observed with the concave mirror but not with the convex mirror. The real image is "real" because the light reflects off the mirror and forms a visible image on the card.

2. The image was smaller and upside down.

Draw Conclusions

3. A concave mirror would be better than a convex mirror. A real image cannot be formed with a convex mirror.

Contents

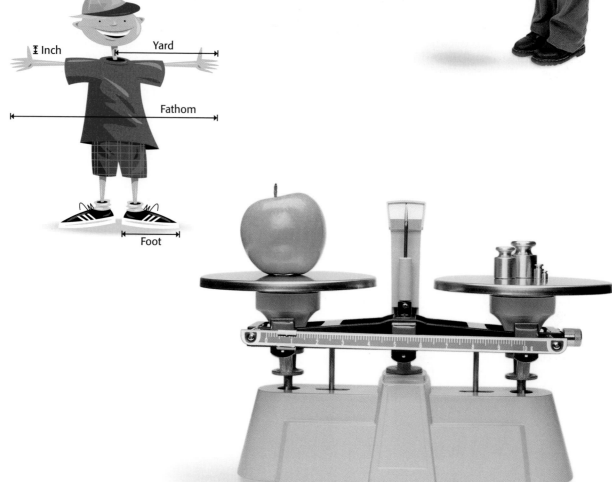

Appendix

✓ Reading Check Answers

Chapter 1   The Energy of Waves
Section 1
Page 4: All waves are disturbances that transmit energy.

Page 6: Electromagnetic waves do not require a medium.

Page 8: A sound wave is a longitudinal wave.

Section 2
Page 11: Shaking the rope faster makes the wavelength shorter; shaking the rope more slowly makes the wavelength longer.

Page 12: 3 Hz

Section 3
Page 15: It refracts.

Page 17: Constructive interference occurs when the crests of one wave overlap the crests of another wave.

Page 18: A standing wave results from a wave that is reflected between two fixed points. Interference from the wave and reflected waves cause certain points to remain at rest and certain points to remain at a large amplitude.

Chapter 2   The Nature of Sound
Section 1
Page 31: Sound waves consist of longitudinal waves carried through a medium.

Page 32: Sound needs a medium in order to travel.

Page 34: Tinnitus is caused by long-term exposure to loud sounds.

Section 2
Page 37: Frequency is the number of crests or troughs made in a given time.

Page 39: The amplitude of a sound increases as the energy of the vibrations that caused the sound increases.

Page 40: An oscilloscope turns sounds into electrical signals and graphs the signals.

Section 3
Page 43: Echolocation helps some animals find food.

Page 44: Sound wave interference can be either constructive or destructive.

Page 46: A standing wave is a pattern of vibration that looks like a wave that is standing still.

Section 4
Page 49: Musical instruments differ in the part of the instrument that vibrates and in the way that the vibrations are made.

Page 51: Music consists of sound waves that have regular patterns, and noise consists of a random mix of frequencies.

Chapter 3   The Nature of Light
Section 1
Page 63: Electric fields can be found around every charged object.

Page 64: The speed of light is about 880,000 times faster than the speed of sound.

Section 2
Page 66: The speed of a wave is determined by multiplying the wavelength and frequency of the wave.

Page 68: Radio waves carry TV signals.

Page 70: White light is the combination of visible light of all wavelengths.

Page 71: Ultraviolet light waves have shorter wavelengths and higher frequencies than visible light waves do.

Page 72: Patients are protected from X rays by special lead-lined aprons.

Section 3
Page 74: The law of reflection states that the angle of incidence equals the angle of reflection.

Page 75: Sample answer: Four light sources are a television screen, a fluorescent light in the classroom, a light bulb, and the tail of a firefly.

Page 76: You can see things outside of a beam of light because light is scattered outside of the beam.

Page 79: The amount that a wave diffracts depends on the wavelength of the wave and the size of the barrier or opening.

Page 80: Constructive interference is interference in which the resulting wave has a greater amplitude than the original waves had.

Section 4
Page 83: Sample answer: Two translucent objects are a frosted window and wax paper.

Page 84: When white light shines on a colored opaque object, some of the colors of light are absorbed and some are reflected.

Page 86: A pigment is a material that gives color to a substance by absorbing some colors of light and reflecting others.

Chapter 4   Light and Our World
Section 1
Page 99: A virtual image is an image through which light does not travel.

Page 100: A concave mirror can be used to make a powerful beam of light by putting a light source at the focal point of the mirror.

Page 102: A convex lens in thicker in the middle than it is at the edges.

Section 2

Page 105: Nearsightedness happens when a person's eye is too long. Farsightedness happens when a person's eye is too short.

Page 106: The three kinds of cones are red, blue, and green.

Section 3

Page 109: When light is coherent, light waves move together as they travel away from their source. Individual waves behave as one wave.

Page 111: Holograms are like photographs because both are images recorded on film.

Page 113: A cordless telephone sends signals by using radio waves.

Page 114: Sample answer: GPS can be used by hikers and campers to find their way in the wilderness. GPS can also be used for treasure hunt games.

Appendix

Study Skills

FoldNote Instructions

Have you ever tried to study for a test or quiz but didn't know where to start? Or have you read a chapter and found that you can remember only a few ideas? Well, FoldNotes are a fun and exciting way to help you learn and remember the ideas you encounter as you learn science!

FoldNotes are tools that you can use to organize concepts. By focusing on a few main concepts, FoldNotes help you learn and remember how the concepts fit together. They can help you see the "big picture." Below you will find instructions for building 10 different FoldNotes.

Pyramid

1. Place a sheet of paper in front of you. Fold the lower left-hand corner of the paper diagonally to the opposite edge of the paper.

2. Cut off the tab of paper created by the fold (at the top).

3. Open the paper so that it is a square. Fold the lower right-hand corner of the paper diagonally to the opposite corner to form a triangle.

4. Open the paper. The creases of the two folds will have created an X.

5. Using scissors, cut along one of the creases. Start from any corner, and stop at the center point to create two flaps. Use tape or glue to attach one of the flaps on top of the other flap.

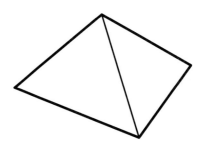

Double Door

1. Fold a sheet of paper in half from the top to the bottom. Then, unfold the paper.

2. Fold the top and bottom edges of the paper to the crease.

Booklet

1. Fold a sheet of paper in half from left to right. Then, unfold the paper.

2. Fold the sheet of paper in half again from the top to the bottom. Then, unfold the paper.

3. Refold the sheet of paper in half from left to right.

4. Fold the top and bottom edges to the center crease.

5. Completely unfold the paper.

6. Refold the paper from top to bottom.

7. Using scissors, cut a slit along the center crease of the sheet from the folded edge to the creases made in step 4. Do not cut the entire sheet in half.

8. Fold the sheet of paper in half from left to right. While holding the bottom and top edges of the paper, push the bottom and top edges together so that the center collapses at the center slit. Fold the four flaps to form a four-page book.

Layered Book

1. Lay one sheet of paper on top of another sheet. Slide the top sheet up so that 2 cm of the bottom sheet is showing.

2. Hold the two sheets together, fold down the top of the two sheets so that you see four 2 cm tabs along the bottom.

3. Using a stapler, staple the top of the FoldNote.

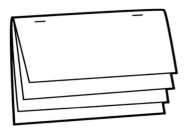

Key-Term Fold

1. Fold a sheet of lined notebook paper in half from left to right.

2. Using scissors, cut along every third line from the right edge of the paper to the center fold to make tabs.

Four-Corner Fold

1. Fold a sheet of paper in half from left to right. Then, unfold the paper.

2. Fold each side of the paper to the crease in the center of the paper.

3. Fold the paper in half from the top to the bottom. Then, unfold the paper.

4. Using scissors, cut the top flap creases made in step 3 to form four flaps.

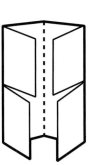

Three-Panel Flip Chart

1. Fold a piece of paper in half from the top to the bottom.

2. Fold the paper in thirds from side to side. Then, unfold the paper so that you can see the three sections.

3. From the top of the paper, cut along each of the vertical fold lines to the fold in the middle of the paper. You will now have three flaps.

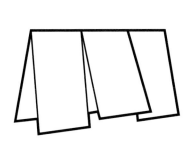

Appendix

Table Fold

1. Fold a piece of paper in half from the top to the bottom. Then, fold the paper in half again.

2. Fold the paper in thirds from side to side.

3. Unfold the paper completely. Carefully trace the fold lines by using a pen or pencil.

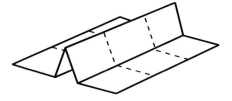

Two-Panel Flip Chart

1. Fold a piece of paper in half from the top to the bottom.

2. Fold the paper in half from side to side. Then, unfold the paper so that you can see the two sections.

3. From the top of the paper, cut along the vertical fold line to the fold in the middle of the paper. You will now have two flaps.

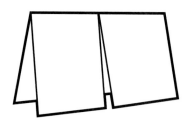

Tri-Fold

1. Fold a piece a paper in thirds from the top to the bottom.

2. Unfold the paper so that you can see the three sections. Then, turn the paper sideways so that the three sections form vertical columns.

3. Trace the fold lines by using a pen or pencil. Label the columns "Know," "Want," and "Learn."

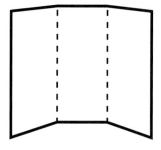

Graphic Organizer Instructions

Have you ever wished that you could "draw out" the many concepts you learn in your science class? Sometimes, being able to *see* how concepts are related really helps you remember what you've learned. Graphic Organizers do just that! They give you a way to draw or map out concepts.

All you need to make a Graphic Organizer is a piece of paper and a pencil. Below you will find instructions for four different Graphic Organizers designed to help you organize the concepts you'll learn in this book.

Spider Map

1. Draw a diagram like the one shown. In the circle, write the main topic.

2. From the circle, draw legs to represent different categories of the main topic. You can have as many categories as you want.

3. From the category legs, draw horizontal lines. As you read the chapter, write details about each category on the horizontal lines.

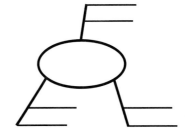

Comparison Table

1. Draw a chart like the one shown. Your chart can have as many columns and rows as you want.

2. In the top row, write the topics that you want to compare.

3. In the left column, write characteristics of the topics that you want to compare. As you read the chapter, fill in the characteristics for each topic in the appropriate boxes.

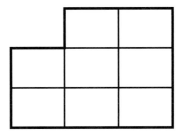

Chain-of-Events-Chart

1. Draw a box. In the box, write the first step of a process or the first event of a timeline.

2. Under the box, draw another box, and use an arrow to connect the two boxes. In the second box, write the next step of the process or the next event in the timeline.

3. Continue adding boxes until the process or timeline is finished.

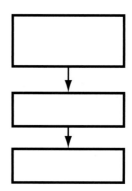

Concept Map

1. Draw a circle in the center of a piece of paper. Write the main idea of the chapter in the center of the circle.

2. From the circle, draw other circles. In those circles, write characteristics of the main idea. Draw arrows from the center circle to the circles that contain the characteristics.

3. From each circle that contains a characteristic, draw other circles. In those circles, write specific details about the characteristic. Draw arrows from each circle that contains a characteristic to the circles that contain specific details. You may draw as many circles as you want.

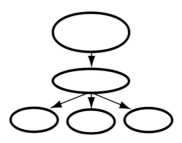

SI Measurement

The International System of Units, or SI, is the standard system of measurement used by many scientists. Using the same standards of measurement makes it easier for scientists to communicate with one another.

SI works by combining prefixes and base units. Each base unit can be used with different prefixes to define smaller and larger quantities. The table below lists common SI prefixes.

SI Prefixes

Prefix	Symbol	Factor	Example
kilo-	k	1,000	kilogram, 1 kg = 1,000 g
hecto-	h	100	hectoliter, 1 hL = 100 L
deka-	da	10	dekameter, 1 dam = 10 m
		1	meter, liter, gram
deci-	d	0.1	decigram, 1 dg = 0.1 g
centi-	c	0.01	centimeter, 1 cm = 0.01 m
milli-	m	0.001	milliliter, 1 mL = 0.001 L
micro-	μ	0.000 001	micrometer, 1 μm = 0.000 001 m

SI Conversion Table

SI units	From SI to English	From English to SI
Length		
kilometer (km) = 1,000 m	1 km = 0.621 mi	1 mi = 1.609 km
meter (m) = 100 cm	1 m = 3.281 ft	1 ft = 0.305 m
centimeter (cm) = 0.01 m	1 cm = 0.394 in.	1 in. = 2.540 cm
millimeter (mm) = 0.001 m	1 mm = 0.039 in.	
micrometer (μm) = 0.000 001 m		
nanometer (nm) = 0.000 000 001 m		
Area		
square kilometer (km^2) = 100 hectares	1 km^2 = 0.386 mi^2	1 mi^2 = 2.590 km^2
hectare (ha) = 10,000 m^2	1 ha = 2.471 acres	1 acre = 0.405 ha
square meter (m^2) = 10,000 cm^2	1 m^2 = 10.764 ft^2	1 ft^2 = 0.093 m^2
square centimeter (cm^2) = 100 mm^2	1 cm^2 = 0.155 in.^{2}	1 in.^{2} = 6.452 cm^2
Volume		
liter (L) = 1,000 mL = 1 dm^3	1 L = 1.057 fl qt	1 fl qt = 0.946 L
milliliter (mL) = 0.001 L = 1 cm^3	1 mL = 0.034 fl oz	1 fl oz = 29.574 mL
microliter (μL) = 0.000 001 L		
Mass		
kilogram (kg) = 1,000 g	1 kg = 2.205 lb	1 lb = 0.454 kg
gram (g) = 1,000 mg	1 g = 0.035 oz	1 oz = 28.350 g
milligram (mg) = 0.001 g		
microgram (μg) = 0.000 001 g		

Temperature Scales

Temperature can be expressed by using three different scales: Fahrenheit, Celsius, and Kelvin. The SI unit for temperature is the kelvin (K).

Although 0 K is much colder than 0°C, a change of 1 K is equal to a change of 1°C.

Three Temperature Scales

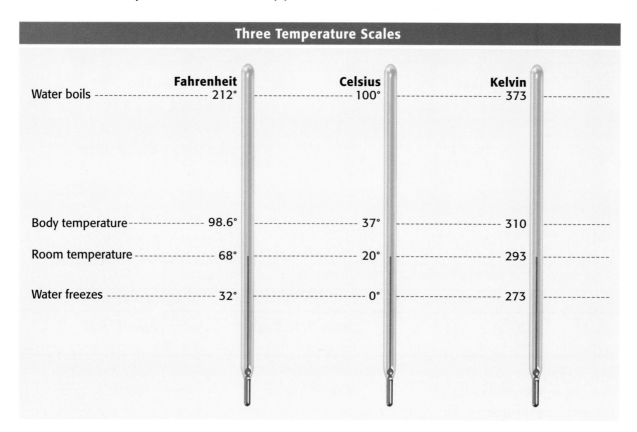

	Fahrenheit	Celsius	Kelvin
Water boils	212°	100°	373
Body temperature	98.6°	37°	310
Room temperature	68°	20°	293
Water freezes	32°	0°	273

Temperature Conversions Table

To convert	Use this equation:	Example
Celsius to Fahrenheit °C → °F	$°F = \left(\dfrac{9}{5} \times °C\right) + 32$	Convert 45°C to °F. $°F = \left(\dfrac{9}{5} \times 45°C\right) + 32 = 113°F$
Fahrenheit to Celsius °F → °C	$°C = \dfrac{5}{9} \times (°F - 32)$	Convert 68°F to °C. $°C = \dfrac{5}{9} \times (68°F - 32) = 20°C$
Celsius to Kelvin °C → K	$K = °C + 273$	Convert 45°C to K. $K = 45°C + 273 = 318\ K$
Kelvin to Celsius K → °C	$°C = K - 273$	Convert 32 K to °C. $°C = 32K - 273 = -241°C$

Measuring Skills

Using a Graduated Cylinder

When using a graduated cylinder to measure volume, keep the following procedures in mind:

1. Place the cylinder on a flat, level surface before measuring liquid.

2. Move your head so that your eye is level with the surface of the liquid.

3. Read the mark closest to the liquid level. On glass graduated cylinders, read the mark closest to the center of the curve in the liquid's surface.

Using a Meterstick or Metric Ruler

When using a meterstick or metric ruler to measure length, keep the following procedures in mind:

1. Place the ruler firmly against the object that you are measuring.

2. Align one edge of the object exactly with the 0 end of the ruler.

3. Look at the other edge of the object to see which of the marks on the ruler is closest to that edge. (Note: Each small slash between the centimeters represents a millimeter, which is one-tenth of a centimeter.)

Using a Triple-Beam Balance

When using a triple-beam balance to measure mass, keep the following procedures in mind:

1. Make sure the balance is on a level surface.

2. Place all of the countermasses at 0. Adjust the balancing knob until the pointer rests at 0.

3. Place the object you wish to measure on the pan. **Caution:** Do not place hot objects or chemicals directly on the balance pan.

4. Move the largest countermass along the beam to the right until it is at the last notch that does not tip the balance. Follow the same procedure with the next-largest countermass. Then, move the smallest countermass until the pointer rests at 0.

5. Add the readings from the three beams together to determine the mass of the object.

6. When determining the mass of crystals or powders, first find the mass of a piece of filter paper. Then, add the crystals or powder to the paper, and remeasure. The actual mass of the crystals or powder is the total mass minus the mass of the paper. When finding the mass of liquids, first find the mass of the empty container. Then, find the combined mass of the liquid and container. The mass of the liquid is the total mass minus the mass of the container.

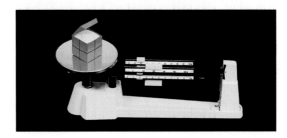

Scientific Methods

The ways in which scientists answer questions and solve problems are called **scientific methods.** The same steps are often used by scientists as they look for answers. However, there is more than one way to use these steps. Scientists may use all of the steps or just some of the steps during an investigation. They may even repeat some of the steps. The goal of using scientific methods is to come up with reliable answers and solutions.

Six Steps of Scientific Methods

1 Ask a Question

Good questions come from careful **observations.** You make observations by using your senses to gather information. Sometimes, you may use instruments, such as microscopes and telescopes, to extend the range of your senses. As you observe the natural world, you will discover that you have many more questions than answers. These questions drive investigations.

Questions beginning with *what, why, how,* and *when* are important in focusing an investigation. Here is an example of a question that could lead to an investigation.

Question: How does acid rain affect plant growth?

2 Form a Hypothesis

After you ask a question, you need to form a **hypothesis.** A hypothesis is a clear statement of what you expect the answer to your question to be. Your hypothesis will represent your best "educated guess" based on what you have observed and what you already know. A good hypothesis is testable. Otherwise, the investigation can go no further. Here is a hypothesis based on the question, "How does acid rain affect plant growth?"

Hypothesis: Acid rain slows plant growth.

The hypothesis can lead to predictions. A prediction is what you think the outcome of your experiment or data collection will be. Predictions are usually stated in an if-then format. Here is a sample prediction for the hypothesis that acid rain slows plant growth.

Prediction: If a plant is watered with only acid rain (which has a pH of 4), then the plant will grow at half its normal rate.

3 Test the Hypothesis

After you have formed a hypothesis and made a prediction, your hypothesis should be tested. One way to test a hypothesis is with a controlled experiment. A **controlled experiment** tests only one factor at a time. In an experiment to test the effect of acid rain on plant growth, the **control group** would be watered with normal rain water. The **experimental group** would be watered with acid rain. All of the plants should receive the same amount of sunlight and water each day. The air temperature should be the same for all groups. However, the acidity of the water will be a variable. In fact, any factor that is different from one group to another is a **variable.** If your hypothesis is correct, then the acidity of the water and plant growth are *dependant variables.* The amount a plant grows is dependent on the acidity of the water. However, the amount of water each plant receives and the amount of sunlight each plant receives are *independent variables.* Either of these factors could change without affecting the other factor.

Sometimes, the nature of an investigation makes a controlled experiment impossible. For example, the Earth's core is surrounded by thousands of meters of rock. Under such circumstances, a hypothesis may be tested by making detailed observations.

4 Analyze the Results

After you have completed your experiments, made your observations, and collected your data, you must analyze all the information you have gathered. Tables and graphs are often used in this step to organize the data.

5 Draw Conclusions

After analyzing your data, you can determine if your results support your hypothesis. If your hypothesis is supported, you (or others) might want to repeat the observations or experiments to verify your results. If your hypothesis is not supported by the data, you may have to check your procedure for errors. You may even have to reject your hypothesis and make a new one. If you cannot draw a conclusion from your results, you may have to try the investigation again or carry out further observations or experiments.

6 Communicate Results

After any scientific investigation, you should report your results. By preparing a written or oral report, you let others know what you have learned. They may repeat your investigation to see if they get the same results. Your report may even lead to another question and then to another investigation.

Scientific Methods in Action

Scientific methods contain loops in which several steps may be repeated over and over again. In some cases, certain steps are unnecessary. Thus, there is not a "straight line" of steps. For example, sometimes scientists find that testing one hypothesis raises new questions and new hypotheses to be tested. And sometimes,

testing the hypothesis leads directly to a conclusion. Furthermore, the steps in scientific methods are not always used in the same order. Follow the steps in the diagram, and see how many different directions scientific methods can take you.

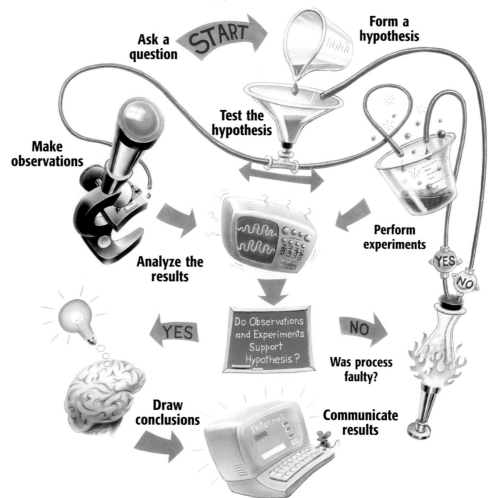

Periodic Table of the Elements

Each square on the table includes an element's name, chemical symbol, atomic number, and atomic mass.

The color of the chemical symbol indicates the physical state at room temperature. Carbon is a solid.

| 6 |
| **C** |
| Carbon |
| 12.0 |

Atomic number — 6
Chemical symbol — C
Element name — Carbon
Atomic mass — 12.0

The background color indicates the type of element. Carbon is a nonmetal.

Period 1

| 1 |
| **H** |
| Hydrogen |
| 1.0 |

Background
Metals
Metalloids
Nonmetals

Chemical symbol
Solid
Liquid
Gas

	Group 1	Group 2	Group 3	Group 4	Group 5	Group 6	Group 7	Group 8	Group 9
Period 2	3 **Li** Lithium 6.9	4 **Be** Beryllium 9.0							
Period 3	11 **Na** Sodium 23.0	12 **Mg** Magnesium 24.3							
Period 4	19 **K** Potassium 39.1	20 **Ca** Calcium 40.1	21 **Sc** Scandium 45.0	22 **Ti** Titanium 47.9	23 **V** Vanadium 50.9	24 **Cr** Chromium 52.0	25 **Mn** Manganese 54.9	26 **Fe** Iron 55.8	27 **Co** Cobalt 58.9
Period 5	37 **Rb** Rubidium 85.5	38 **Sr** Strontium 87.6	39 **Y** Yttrium 88.9	40 **Zr** Zirconium 91.2	41 **Nb** Niobium 92.9	42 **Mo** Molybdenum 95.9	43 **Tc** Technetium (98)	44 **Ru** Ruthenium 101.1	45 **Rh** Rhodium 102.9
Period 6	55 **Cs** Cesium 132.9	56 **Ba** Barium 137.3	57 **La** Lanthanum 138.9	72 **Hf** Hafnium 178.5	73 **Ta** Tantalum 180.9	74 **W** Tungsten 183.8	75 **Re** Rhenium 186.2	76 **Os** Osmium 190.2	77 **Ir** Iridium 192.2
Period 7	87 **Fr** Francium (223)	88 **Ra** Radium (226)	89 **Ac** Actinium (227)	104 **Rf** Rutherfordium (261)	105 **Db** Dubnium (262)	106 **Sg** Seaborgium (263)	107 **Bh** Bohrium (264)	108 **Hs** Hassium (265)†	109 **Mt** Meitnerium (268)†

† Estimated from currently available IUPAC data.

A row of elements is called a *period*.

A column of elements is called a *group* or *family*.

Values in parentheses are of the most stable isotope of the element.

These elements are placed below the table to allow the table to be narrower.

Lanthanides	58 **Ce** Cerium 140.1	59 **Pr** Praseodymium 140.9	60 **Nd** Neodymium 144.2	61 **Pm** Promethium (145)	62 **Sm** Samarium 150.4
Actinides	90 **Th** Thorium 232.0	91 **Pa** Protactinium 231.0	92 **U** Uranium 238.0	93 **Np** Neptunium (237)	94 **Pu** Plutonium (244)

Topic: **Periodic Table**
Go To: **go.hrw.com**
Keyword: **HN0 PERIODIC**
Visit the HRW Web site for updates on the periodic table.

This zigzag line reminds you where the metals, nonmetals, and metalloids are.

The names and three-letter symbols of elements are temporary. They are based on the atomic numbers of the elements. Official names and symbols will be approved by an international committee of scientists.

Group 18
2 **He** Helium 4.0

Group 13	Group 14	Group 15	Group 16	Group 17
5 **B** Boron 10.8	6 **C** Carbon 12.0	7 **N** Nitrogen 14.0	8 **O** Oxygen 16.0	9 **F** Fluorine 19.0	10 **Ne** Neon 20.2
13 **Al** Aluminum 27.0	14 **Si** Silicon 28.1	15 **P** Phosphorus 31.0	16 **S** Sulfur 32.1	17 **Cl** Chlorine 35.5	18 **Ar** Argon 39.9

Group 10	Group 11	Group 12						
28 **Ni** Nickel 58.7	29 **Cu** Copper 63.5	30 **Zn** Zinc 65.4	31 **Ga** Gallium 69.7	32 **Ge** Germanium 72.6	33 **As** Arsenic 74.9	34 **Se** Selenium 79.0	35 **Br** Bromine 79.9	36 **Kr** Krypton 83.8
46 **Pd** Palladium 106.4	47 **Ag** Silver 107.9	48 **Cd** Cadmium 112.4	49 **In** Indium 114.8	50 **Sn** Tin 118.7	51 **Sb** Antimony 121.8	52 **Te** Tellurium 127.6	53 **I** Iodine 126.9	54 **Xe** Xenon 131.3
78 **Pt** Platinum 195.1	79 **Au** Gold 197.0	80 **Hg** Mercury 200.6	81 **Tl** Thallium 204.4	82 **Pb** Lead 207.2	83 **Bi** Bismuth 209.0	84 **Po** Polonium (209)	85 **At** Astatine (210)	86 **Rn** Radon (222)
110 **Ds** Darmstadtium (269)[†]	111 **Uuu** Unununium (272)[†]	112 **Uub** Ununbium (277)[†]		114 **Uuq** Ununquadium (285)[†]				

63 **Eu** Europium 152.0	64 **Gd** Gadolinium 157.2	65 **Tb** Terbium 158.9	66 **Dy** Dysprosium 162.5	67 **Ho** Holmium 164.9	68 **Er** Erbium 167.3	69 **Tm** Thulium 168.9	70 **Yb** Ytterbium 173.0	71 **Lu** Lutetium 175.0
95 **Am** Americium (243)	96 **Cm** Curium (247)	97 **Bk** Berkelium (247)	98 **Cf** Californium (251)	99 **Es** Einsteinium (252)	100 **Fm** Fermium (257)	101 **Md** Mendelevium (258)	102 **No** Nobelium (259)	103 **Lr** Lawrencium (262)

Appendix

Making Charts and Graphs

Pie Charts

A pie chart shows how each group of data relates to all of the data. Each part of the circle forming the chart represents a category of the data. The entire circle represents all of the data. For example, a biologist studying a hardwood forest in Wisconsin found that there were five different types of trees. The data table at right summarizes the biologist's findings.

Wisconsin Hardwood Trees	
Type of tree	Number found
Oak	600
Maple	750
Beech	300
Birch	1,200
Hickory	150
Total	3,000

How to Make a Pie Chart

1 To make a pie chart of these data, first find the percentage of each type of tree. Divide the number of trees of each type by the total number of trees, and multiply by 100.

$$\frac{600 \text{ oak}}{3,000 \text{ trees}} \times 100 = 20\%$$

$$\frac{750 \text{ maple}}{3,000 \text{ trees}} \times 100 = 25\%$$

$$\frac{300 \text{ beech}}{3,000 \text{ trees}} \times 100 = 10\%$$

$$\frac{1,200 \text{ birch}}{3,000 \text{ trees}} \times 100 = 40\%$$

$$\frac{150 \text{ hickory}}{3,000 \text{ trees}} \times 100 = 5\%$$

2 Now, determine the size of the wedges that make up the pie chart. Multiply each percentage by 360°. Remember that a circle contains 360°.

20% × 360° = 72°   25% × 360° = 90°

10% × 360° = 36°   40% × 360° = 144°

5% × 360° = 18°

3 Check that the sum of the percentages is 100 and the sum of the degrees is 360.

20% + 25% + 10% + 40% + 5% = 100%

72° + 90° + 36° + 144° + 18° = 360°

4 Use a compass to draw a circle and mark the center of the circle.

5 Then, use a protractor to draw angles of 72°, 90°, 36°, 144°, and 18° in the circle.

6 Finally, label each part of the chart, and choose an appropriate title.

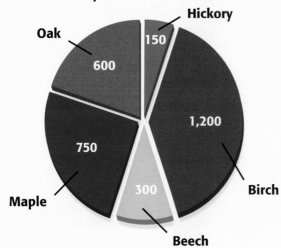

A Community of Wisconsin Hardwood Trees

Line Graphs

Line graphs are most often used to demonstrate continuous change. For example, Mr. Smith's students analyzed the population records for their hometown, Appleton, between 1900 and 2000. Examine the data at right.

Because the year and the population change, they are the *variables*. The population is determined by, or dependent on, the year. Therefore, the population is called the **dependent variable,** and the year is called the **independent variable.** Each set of data is called a **data pair.** To prepare a line graph, you must first organize data pairs into a table like the one at right.

Population of Appleton, 1900–2000	
Year	**Population**
1900	1,800
1920	2,500
1940	3,200
1960	3,900
1980	4,600
2000	5,300

How to Make a Line Graph

❶ Place the independent variable along the horizontal (*x*) axis. Place the dependent variable along the vertical (*y*) axis.

❷ Label the *x*-axis "Year" and the *y*-axis "Population." Look at your largest and smallest values for the population. For the *y*-axis, determine a scale that will provide enough space to show these values. You must use the same scale for the entire length of the axis. Next, find an appropriate scale for the *x*-axis.

❸ Choose reasonable starting points for each axis.

❹ Plot the data pairs as accurately as possible.

❺ Choose a title that accurately represents the data.

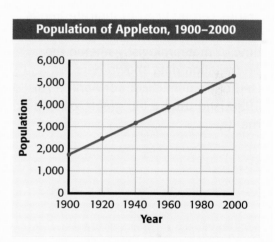

How to Determine Slope

Slope is the ratio of the change in the *y*-value to the change in the *x*-value, or "rise over run."

❶ Choose two points on the line graph. For example, the population of Appleton in 2000 was 5,300 people. Therefore, you can define point *a* as (2000, 5,300). In 1900, the population was 1,800 people. You can define point *b* as (1900, 1,800).

❷ Find the change in the *y*-value. (*y* at point *a*) − (*y* at point *b*) = 5,300 people − 1,800 people = 3,500 people

❸ Find the change in the *x*-value. (*x* at point *a*) − (*x* at point *b*) = 2000 − 1900 = 100 years

❹ Calculate the slope of the graph by dividing the change in *y* by the change in *x*.

$$slope = \frac{change\ in\ y}{change\ in\ x}$$

$$slope = \frac{3,500\ people}{100\ years}$$

$$slope = 35\ people\ per\ year$$

In this example, the population in Appleton increased by a fixed amount each year. The graph of these data is a straight line. Therefore, the relationship is **linear.** When the graph of a set of data is not a straight line, the relationship is **nonlinear.**

Appendix

Using Algebra to Determine Slope

The equation in step 4 may also be arranged to be

$$y = kx$$

where y represents the change in the y-value, k represents the slope, and x represents the change in the x-value.

$$slope = \frac{change\ in\ y}{change\ in\ x}$$

$$k = \frac{y}{x}$$

$$k \times x = \frac{y \times x}{x}$$

$$kx = y$$

Bar Graphs

Bar graphs are used to demonstrate change that is not continuous. These graphs can be used to indicate trends when the data cover a long period of time. A meteorologist gathered the precipitation data shown here for Hartford, Connecticut, for April 1–15, 1996, and used a bar graph to represent the data.

Precipitation in Hartford, Connecticut April 1–15, 1996			
Date	Precipitation (cm)	Date	Precipitation (cm)
April 1	0.5	April 9	0.25
April 2	1.25	April 10	0.0
April 3	0.0	April 11	1.0
April 4	0.0	April 12	0.0
April 5	0.0	April 13	0.25
April 6	0.0	April 14	0.0
April 7	0.0	April 15	6.50
April 8	1.75		

How to Make a Bar Graph

1 Use an appropriate scale and a reasonable starting point for each axis.

2 Label the axes, and plot the data.

3 Choose a title that accurately represents the data.

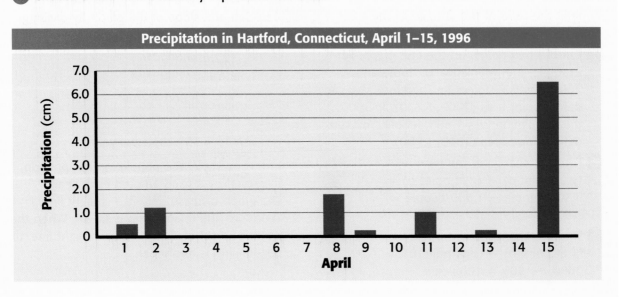

Precipitation in Hartford, Connecticut, April 1–15, 1996

Math Refresher

Science requires an understanding of many math concepts. The following pages will help you review some important math skills.

Averages

An **average,** or **mean,** simplifies a set of numbers into a single number that *approximates* the value of the set.

> **Example:** Find the average of the following set of numbers: 5, 4, 7, and 8.

Step 1: Find the sum.

$$5 + 4 + 7 + 8 = 24$$

Step 2: Divide the sum by the number of numbers in your set. Because there are four numbers in this example, divide the sum by 4.

$$\frac{24}{4} = 6$$

The average, or mean, is **6.**

Ratios

A **ratio** is a comparison between numbers, and it is usually written as a fraction.

> **Example:** Find the ratio of thermometers to students if you have 36 thermometers and 48 students in your class.

Step 1: Make the ratio.

$$\frac{36 \text{ thermometers}}{48 \text{ students}}$$

Step 2: Reduce the fraction to its simplest form.

$$\frac{36}{48} = \frac{36 \div 12}{48 \div 12} = \frac{3}{4}$$

The ratio of thermometers to students is **3 to 4,** or $\frac{3}{4}$. The ratio may also be written in the form 3:4.

Proportions

A **proportion** is an equation that states that two ratios are equal.

$$\frac{3}{1} = \frac{12}{4}$$

To solve a proportion, first multiply across the equal sign. This is called *cross-multiplication.* If you know three of the quantities in a proportion, you can use cross-multiplication to find the fourth.

> **Example:** Imagine that you are making a scale model of the solar system for your science project. The diameter of Jupiter is 11.2 times the diameter of the Earth. If you are using a plastic-foam ball that has a diameter of 2 cm to represent the Earth, what must the diameter of the ball representing Jupiter be?

$$\frac{11.2}{1} = \frac{x}{2 \text{ cm}}$$

Step 1: Cross-multiply.

$$\frac{11.2}{1} \diagdown\!\!\!\!\!\diagup \frac{x}{2}$$

$$11.2 \times 2 = x \times 1$$

Step 2: Multiply.

$$22.4 = x \times 1$$

Step 3: Isolate the variable by dividing both sides by 1.

$$x = \frac{22.4}{1}$$

$$x = 22.4 \text{ cm}$$

You will need to use a ball that has a diameter of **22.4** cm to represent Jupiter.

Percentages

A **percentage** is a ratio of a given number to 100.

> **Example:** What is 85% of 40?

Step 1: Rewrite the percentage by moving the decimal point two places to the left.

$$0.\underset{\smile}{85}$$

Step 2: Multiply the decimal by the number that you are calculating the percentage of.

$$0.85 \times 40 = 34$$

85% of 40 is **34.**

Decimals

To **add** or **subtract decimals,** line up the digits vertically so that the decimal points line up. Then, add or subtract the columns from right to left. Carry or borrow numbers as necessary.

> **Example:** Add the following numbers: 3.1415 and 2.96.

Step 1: Line up the digits vertically so that the decimal points line up.

$$
\begin{array}{r}
3.1415 \\
+\ 2.96 \\
\hline
\end{array}
$$

Step 2: Add the columns from right to left, and carry when necessary.

$$
\begin{array}{r}
{}^{1}\ {}^{1}\ \ \ \ \\
3.1415 \\
+\ 2.96 \\
\hline
6.1015
\end{array}
$$

The sum is **6.1015.**

Fractions

Numbers tell you how many; **fractions** tell you *how much of a whole*.

> **Example:** Your class has 24 plants. Your teacher instructs you to put 5 plants in a shady spot. What fraction of the plants in your class will you put in a shady spot?

Step 1: In the denominator, write the total number of parts in the whole.

$$\frac{?}{24}$$

Step 2: In the numerator, write the number of parts of the whole that are being considered.

$$\frac{5}{24}$$

So, $\frac{5}{24}$ of the plants will be in the shade.

Reducing Fractions

It is usually best to express a fraction in its simplest form. Expressing a fraction in its simplest form is called *reducing* a fraction.

> **Example:** Reduce the fraction $\frac{30}{45}$ to its simplest form.

Step 1: Find the largest whole number that will divide evenly into both the numerator and denominator. This number is called the *greatest common factor* (GCF).

Factors of the numerator 30:
    1, 2, 3, 5, 6, 10, **15,** 30

Factors of the denominator 45:
    1, 3, 5, 9, **15,** 45

Step 2: Divide both the numerator and the denominator by the GCF, which in this case is 15.

$$\frac{30}{45} = \frac{30 \div 15}{45 \div 15} = \frac{2}{3}$$

Thus, $\frac{30}{45}$ reduced to its simplest form is $\frac{2}{3}$.

Adding and Subtracting Fractions

To **add** or **subtract fractions** that have the **same denominator,** simply add or subtract the numerators.

Examples:

$$\frac{3}{5} + \frac{1}{5} = ? \text{ and } \frac{3}{4} - \frac{1}{4} = ?$$

Step 1: Add or subtract the numerators.

$$\frac{3}{5} + \frac{1}{5} = \frac{4}{} \text{ and } \frac{3}{4} - \frac{1}{4} = \frac{2}{}$$

Step 2: Write the sum or difference over the denominator.

$$\frac{3}{5} + \frac{1}{5} = \frac{4}{5} \text{ and } \frac{3}{4} - \frac{1}{4} = \frac{2}{4}$$

Step 3: If necessary, reduce the fraction to its simplest form.

$$\frac{4}{5} \text{ cannot be reduced, and } \frac{2}{4} = \frac{1}{2}.$$

To **add** or **subtract fractions** that have **different denominators,** first find the least common denominator (LCD).

Examples:

$$\frac{1}{2} + \frac{1}{6} = ? \text{ and } \frac{3}{4} - \frac{2}{3} = ?$$

Step 1: Write the equivalent fractions that have a common denominator.

$$\frac{3}{6} + \frac{1}{6} = ? \text{ and } \frac{9}{12} - \frac{8}{12} = ?$$

Step 2: Add or subtract the fractions.

$$\frac{3}{6} + \frac{1}{6} = \frac{4}{6} \text{ and } \frac{9}{12} - \frac{8}{12} = \frac{1}{12}$$

Step 3: If necessary, reduce the fraction to its simplest form.

The fraction $\frac{4}{6} = \frac{2}{3}$, and $\frac{1}{12}$ cannot be reduced.

Multiplying Fractions

To **multiply fractions,** multiply the numerators and the denominators together, and then reduce the fraction to its simplest form.

Example:

$$\frac{5}{9} \times \frac{7}{10} = ?$$

Step 1: Multiply the numerators and denominators.

$$\frac{5}{9} \times \frac{7}{10} = \frac{5 \times 7}{9 \times 10} = \frac{35}{90}$$

Step 2: Reduce the fraction.

$$\frac{35}{90} = \frac{35 \div 5}{90 \div 5} = \frac{7}{18}$$

Dividing Fractions

To **divide fractions,** first rewrite the divisor (the number you divide by) upside down. This number is called the *reciprocal* of the divisor. Then multiply and reduce if necessary.

Example:

$$\frac{5}{8} \div \frac{3}{2} = ?$$

Step 1: Rewrite the divisor as its reciprocal.

$$\frac{3}{2} \rightarrow \frac{2}{3}$$

Step 2: Multiply the fractions.

$$\frac{5}{8} \times \frac{2}{3} = \frac{5 \times 2}{8 \times 3} = \frac{10}{24}$$

Step 3: Reduce the fraction.

$$\frac{10}{24} = \frac{10 \div 2}{24 \div 2} = \frac{5}{12}$$

Appendix

Scientific Notation

Scientific notation is a short way of representing very large and very small numbers without writing all of the place-holding zeros.

> **Example:** Write 653,000,000 in scientific notation.

Step 1: Write the number without the place-holding zeros.

653

Step 2: Place the decimal point after the first digit.

6.53

Step 3: Find the exponent by counting the number of places that you moved the decimal point.

6.53000000

The decimal point was moved eight places to the left. Therefore, the exponent of 10 is positive 8. If you had moved the decimal point to the right, the exponent would be negative.

Step 4: Write the number in scientific notation.

$$6.53 \times 10^8$$

Area

Area is the number of square units needed to cover the surface of an object.

Formulas:

area of a square = side × side
area of a rectangle = length × width
area of a triangle = $\frac{1}{2}$ × base × height

Examples: Find the areas.

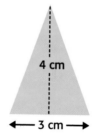

Triangle

$area = \frac{1}{2} \times base \times height$

$area = \frac{1}{2} \times 3\ cm \times 4\ cm$

$area = \textbf{6 cm}^2$

4 cm

3 cm

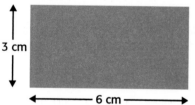

3 cm

6 cm

Rectangle

area = length × width
area = 6 cm × 3 cm
area = **18 cm²**

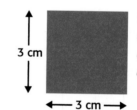

3 cm

3 cm

Square

area = side × side
area = 3 cm × 3 cm
area = **9 cm²**

Volume

Volume is the amount of space that something occupies.

Formulas:

volume of a cube =
side × side × side

volume of a prism =
area of base × height

Examples:

Find the volume of the solids.

Cube

volume = side × side × side
volume = 4 cm × 4 cm × 4 cm
volume = **64 cm³**

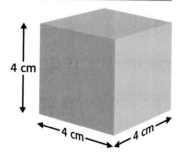

4 cm

4 cm

4 cm

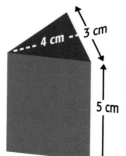

4 cm

3 cm

5 cm

Prism

volume = area of base × height
volume = (area of triangle) × height
volume = ($\frac{1}{2}$ × 3 cm × 4 cm) × 5 cm
volume = 6 cm² × 5 cm
volume = **30 cm³**

Physical Science Laws and Principles

Law of Conservation of Energy

> **The law of conservation of energy states that energy can be neither created nor destroyed.**

The total amount of energy in a closed system is always the same. Energy can be changed from one form to another, but all of the different forms of energy in a system always add up to the same total amount of energy no matter how many energy conversions occur.

Law of Universal Gravitation

> **The law of universal gravitation states that all objects in the universe attract each other by a force called *gravity*. The size of the force depends on the masses of the objects and the distance between objects.**

The first part of the law explains why a bowling ball is much harder to lift than a table-tennis ball. Because the bowling ball has a much larger mass than the table-tennis ball does, the amount of gravity between the Earth and the bowling ball is greater than the amount of gravity between the Earth and the table-tennis ball.

The second part of the law explains why a satellite can remain in orbit around the Earth. The satellite is carefully placed at a distance great enough to prevent the Earth's gravity from immediately pulling the satellite down but small enough to prevent the satellite from completely escaping the Earth's gravity and wandering off into space.

Newton's Laws of Motion

> **Newton's first law of motion states that an object at rest remains at rest and an object in motion remains in motion at constant speed and in a straight line unless acted on by an unbalanced force.**

The first part of the law explains why a football will remain on a tee until it is kicked off or until a gust of wind blows it off.

The second part of the law explains why a bike rider will continue moving forward after the bike comes to an abrupt stop. Gravity and the friction of the sidewalk will eventually stop the rider.

> **Newton's second law of motion states that the acceleration of an object depends on the mass of the object and the amount of force applied.**

The first part of the law explains why the acceleration of a 4 kg bowling ball will be greater than the acceleration of a 6 kg bowling ball if the same force is applied to both.

The second part of the law explains why the acceleration of a bowling ball will be larger if a larger force is applied to the bowling ball.

The relationship of acceleration (a) to mass (m) and force (F) can be expressed mathematically by the following equation:

$$acceleration = \frac{force}{mass}, \text{ or } a = \frac{F}{m}$$

This equation is often rearranged to the form

$$force = mass \times acceleration$$
$$\text{or}$$
$$F = m \times a$$

> **Newton's third law of motion states that whenever one object exerts a force on a second object, the second object exerts an equal and opposite force on the first.**

This law explains that a runner is able to move forward because of the equal and opposite force that the ground exerts on the runner's foot after each step.

Law of Reflection

The law of reflection states that the angle of incidence is equal to the angle of reflection. This law explains why light reflects off a surface at the same angle that the light strikes the surface.

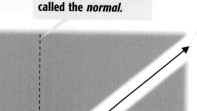

The beam of light traveling toward the mirror is called the *incident beam.*

A line perpendicular to the mirror's surface is called the *normal.*

The beam of light reflected off the mirror is called the *reflected beam.*

The angle between the incident beam and the normal is called the *angle of incidence.*

The angle between the reflected beam and the normal is called the *angle of reflection.*

Charles's Law

Charles's law states that for a fixed amount of gas at a constant pressure, the volume of the gas increases as the temperature of the gas increases. Likewise, the volume of the gas decreases as the temperature of the gas decreases.

If a basketball that was inflated indoors is left outside on a cold winter day, the air particles inside the ball will move more slowly. They will hit the sides of the basketball less often and with less force. The ball will get smaller as the volume of the air decreases.

Boyle's Law

Boyle's law states that for a fixed amount of gas at a constant temperature, the volume of a gas increases as the pressure of the gas decreases. Likewise, the volume of a gas decreases as its pressure increases.

If an inflated balloon is pulled down to the bottom of a swimming pool, the pressure of the water on the balloon increases. The pressure of the air particles inside the balloon must increase to match that of the water outside, so the volume of the air inside the balloon decreases.

Pascal's Principle

Pascal's principle states that a change in pressure at any point in an enclosed fluid will be transmitted equally to all parts of that fluid.

When a mechanic uses a hydraulic jack to raise an automobile off the ground, he or she increases the pressure on the fluid in the jack by pushing on the jack handle. The pressure is transmitted equally to all parts of the fluid-filled jacking system. As fluid presses the jack plate against the frame of the car, the car is lifed off the ground.

Archimedes' Principle

Archimedes' principle states that the buoyant force on an object in a fluid is equal to the weight of the volume of fluid that the object displaces.

A person floating in a swimming pool displaces 20 L of water. The weight of that volume of water is about 200 N. Therefore, the buoyant force on the person is 200 N.

Bernoulli's Principle

Bernoulli's principle states that as the speed of a moving fluid increases, the fluid's pressure decreases.

The lift on an airplane wing or on a Frisbee® can be explained in part by using Bernoulli's principle. Because of the shape of the Frisbee, the air moving over the top of the Frisbee must travel farther than the air below the Frisbee in the same amount of time. In other words, the air above the Frisbee is moving faster than the air below it. This faster-moving air above the Frisbee exerts less pressure than the slower-moving air below it does. The resulting increased pressure below exerts an upward force and pushes the Frisbee up.

Useful Equations

Average speed

$$average\ speed = \frac{total\ distance}{total\ time}$$

Example: A bicycle messenger traveled a distance of 136 km in 8 h. What was the messenger's average speed?

$$\frac{136\ km}{8\ h} = 17\ km/h$$

The messenger's average speed was **17 km/h.**

Average acceleration

$$\frac{average}{acceleration} = \frac{final\ velocity - starting\ velocity}{time\ it\ takes\ to\ change\ velocity}$$

Example: Calculate the average acceleration of an Olympic 100 m dash sprinter who reaches a velocity of 20 m/s south at the finish line. The race was in a straight line and lasted 10 s.

$$\frac{20\ m/s - 0\ m/s}{10s} = 2\ m/s/s$$

The sprinter's average acceleration is **2 m/s/s south.**

Net force

Forces in the Same Direction

When forces are in the same direction, add the forces together to determine the net force.

Example: Calculate the net force on a stalled car that is being pushed by two people. One person is pushing with a force of 13 N northwest, and the other person is pushing with a force of 8 N in the same direction.

$$13\ N + 8\ N = 21\ N$$

The net force is **21 N northwest.**

Forces in Opposite Directions

When forces are in opposite directions, subtract the smaller force from the larger force to determine the net force. The net force will be in the direction of the larger force.

Example: Calculate the net force on a rope that is being pulled on each end. One person is pulling on one end of the rope with a force of 12 N south. Another person is pulling on the opposite end of the rope with a force of 7 N north.

$$12\ N - 7\ N = 5\ N$$

The net force is **5 N south.**

Work

Work is done by exerting a force through a distance. Work has units of joules (J), which are equivalent to Newton-meters.

$$Work = F \times d$$

> **Example:** Calculate the amount of work done by a man who lifts a 100 N toddler 1.5 m off the floor.
>
> $Work$ = 100 N × 1.5 m = 150 N•m = 150 J
>
> The man did **150 J** of work.

Power

Power is the rate at which work is done. Power is measured in watts (W), which are equivalent to joules per second.

$$P = \frac{Work}{t}$$

> **Example:** Calculate the power of a weightlifter who raises a 300 N barbell 2.1 m off the floor in 1.25 s.
>
> $Work$ = 300 N × 2.1 m = 630 N•m = 630 J
>
> $P = \dfrac{630 \text{ J}}{1.25 \text{ s}} = \dfrac{504 \text{ J}}{\text{s}} = 504 \text{ W}$
>
> The weightlifter has **504 W** of power.

Pressure

Pressure is the force exerted over a given area. The SI unit for pressure is the pascal (Pa).

$$pressure = \frac{force}{area}$$

> **Example:** Calculate the pressure of the air in a soccer ball if the air exerts a force of 25,000 N over an area of 0.15 m².
>
> $pressure = \dfrac{25,000 \text{ N}}{0.15 \text{ m}^2} = \dfrac{167,000 \text{ N}}{\text{m}^2} = 167,000 \text{ Pa}$
>
> The pressure of the air inside the soccer ball is **167,000 Pa.**

Density

$$density = \frac{mass}{volume}$$

> **Example:** Calculate the density of a sponge that has a mass of 10 g and a volume of 40 cm³.
>
> $\dfrac{10 \text{ g}}{40 \text{ cm}^3} = \dfrac{0.25 \text{ g}}{\text{cm}^3}$
>
> The density of the sponge is $\dfrac{0.25 \text{ g}}{\text{cm}^3}$.

Concentration

$$concentration = \frac{mass \; of \; solute}{volume \; of \; solvent}$$

> **Example:** Calculate the concentration of a solution in which 10 g of sugar is dissolved in 125 mL of water.
>
> $\dfrac{10 \text{ g of sugar}}{125 \text{ mL of water}} = \dfrac{0.08 \text{ g}}{\text{mL}}$
>
> The concentration of this solution is $\dfrac{0.08 \text{ g}}{\text{mL}}$.

Glossary

A

absorption in optics, the transfer of light energy to particles of matter (76)

amplitude the maximum distance that the particles of a wave's medium vibrate from their rest position (10)

C

concave lens a lens that is thinner in the middle than at the edges (103)

concave mirror a mirror that is curved inward like the inside of a spoon (100)

convex lens a lens that is thicker in the middle than at the edges (102)

convex mirror a mirror that is curved outward like the back of a spoon (101)

D

decibel the most common unit used to measure loudness (symbol, dB) (40)

diffraction a change in the direction of a wave when the wave finds an obstacle or an edge, such as an opening (16, 79)

Doppler effect an observed change in the frequency of a wave when the source or observer is moving (38)

E

echo a reflected sound wave (42)

echolocation the process of using reflected sound waves to find objects; used by animals such as bats (43)

electromagnetic spectrum all of the frequencies or wavelengths of electromagnetic radiation (66)

electromagnetic wave a wave that consists of electric and magnetic fields that vibrate at right angles to each other (62)

F

farsightedness a condition in which the lens of the eye focuses distant objects behind rather than on the retina (105)

frequency the number of waves produced in a given amount of time (12)

H

hologram a piece of film that produces a three-dimensional image of an object; made by using laser light (111)

I

interference the combination of two or more waves that results in a single wave (17, 44, 80)

L

laser a device that produces intense light of only one wavelength and color (109)

lens a transparent object that refracts light waves such that they converge or diverge to create an image (102)

longitudinal wave a wave in which the particles of the medium vibrate parallel to the direction of wave motion (8)

loudness the extent to which a sound can be heard (39)

M

medium a physical environment in which phenomena occur (5, 32)

N

nearsightedness a condition in which the lens of the eye focuses distant objects in front of rather than on the retina (105)

noise a sound that consists of a random mix of frequencies (51)

O

opaque (oh PAYK) describes an object that is not transparent or translucent (83)

P

pigment a substance that gives another substance or a mixture its color (86)

pitch a measure of how high or low a sound is perceived to be, depending on the frequency of the sound wave (37)

plane mirror a mirror that has a flat surface (99)

R

radiation the transfer of energy as electromagnetic waves (63)

reflection the bouncing back of a ray of light, sound, or heat when the ray hits a surface that it does not go through (14, 74)

refraction the bending of a wave as the wave passes between two substances in which the speed of the wave differs (15, 77)

resonance a phenomenon that occurs when two objects naturally vibrate at the same frequency; the sound produced by one object causes the other object to vibrate (19, 46)

S

scattering an interaction of light with matter that causes light to change its energy, direction of motion, or both (76)

sonic boom the explosive sound heard when a shock wave from an object traveling faster than the speed of sound reaches a person's ears (45)

sound quality the result of the blending of several pitches through interference (48)

sound wave a longitudinal wave that is caused by vibrations and that travels through a material medium (31)

standing wave a pattern of vibration that simulates a wave that is standing still (18, 46)

T

translucent (trans LOO suhnt) describes matter that transmits light but that does not transmit an image (83)

transmission the passing of light or other form of energy through matter (82)

transparent describes matter that allows light to pass through with little interference (83)

transverse wave a wave in which the particles of the medium move perpendicularly to the direction the wave is traveling (7)

W

wave a periodic disturbance in a solid, liquid, or gas as energy is transmitted through a medium (4)

wavelength the distance from any point on a wave to an identical point on the next wave (11)

wave speed the speed at which a wave travels through a medium (12)

Spanish Glossary

A

absorption/absorción en la óptica, la transferencia de energía luminosa a las partículas de materia (76)

amplitude/amplitud la distancia máxima a la que vibran las partículas del medio de una onda a partir de su posición de reposo (10)

C

concave lens/lente cóncava una lente que es más delgada en la parte media que en los bordes (103)

concave mirror/espejo cóncavo un espejo que está curvado hacia adentro como la parte interior de una cuchara (100)

convex lens/lente convexa una lente que es más gruesa en la parte media que en los bordes (102)

convex mirror/espejo convexo un espejo que está curvado hacia fuera como la parte de atrás de una cuchara (101)

D

decibel/decibel la unidad más común que se usa para medir el volumen del sonido (símbolo: dB) (40)

diffraction/difracción un cambio en la dirección de una onda cuando ésta se encuentra con un obstáculo o un borde, tal como una abertura (16, 79)

Doppler effect/efecto Doppler un cambio que se observa en la frecuencia de una onda cuando la fuente o el observador está en movimiento (38)

E

echo/eco una onda de sonido reflejada (42)

echolocation/ecolocación el proceso de usar ondas de sonido reflejadas para buscar objetos; utilizado por animales tales como los murciélagos (43)

electromagnetic spectrum/espectro electromagnético todas las frecuencias o longitudes de onda de la radiación electromagnética (66)

electromagnetic wave/onda electromagnética una onda que está formada por campos eléctricos y magnéticos que vibran formando un ángulo recto unos con otros (62)

F

farsightedness/hipermetropía condición en la que el cristalino del ojo enfoca los objetos lejanos detrás de la retina en lugar de en ella (105)

frequency/frecuencia el número de ondas producidas en una cantidad de tiempo determinada (12)

H

hologram/holograma una porción de película que produce una imagen tridimensional de un objeto mediante luz láser (111)

I

interference/interferencia la combinación de dos o más ondas que resulta en una sola onda (17, 44, 80)

L

laser/láser un aparato que produce una luz intensa de únicamente una longitud de onda y color (109)

lens/lente un objeto transparente que refracta las ondas de luz de modo que converjan o diverjan para crear una imagen (102)

longitudinal wave/onda longitudinal una onda en la que las partículas del medio vibran paralelamente a la dirección del movimiento de la onda (8)

loudness/volumen el grado al que se escucha un sonido (39)

M

medium/medio un ambiente físico en el que ocurren fenómenos (5, 32)

N

nearsightedness/miopía condición en la que el cristalino del ojo enfoca los objetos lejanos delante de la retina en lugar de en ella (105)

noise/ruido un sonido que está constituido por una mezcla aleatoria de frecuencias (51)

O

opaque/opaco término que describe un objeto que no es transparente ni translúcido (83)

P

pigment/pigmento una substancia que le da color a otra substancia o mezcla (86)

pitch/altura tonal una medida de qué tan agudo o grave se percibe un sonido, dependiendo de la frecuencia de la onda sonora (37)

plane mirror/espejo plano un espejo que tiene una superficie plana (99)

R

radiation/radiación la transferencia de energía en forma de ondas electromagnéticas (63)

reflection/reflexión el rebote de un rayo de luz, sonido o calor cuando el rayo golpea una superficie pero no la atraviesa (14, 74)

refraction/refracción el curvamiento de una onda cuando ésta pasa entre dos substancias en las que su velocidad difiere (15, 77)

resonance/resonancia un fenómeno que ocurre cuando dos objetos vibran naturalmente a la misma frecuencia; el sonido producido por un objeto hace que el otro objeto vibre (19, 46)

S

scattering/dispersión una interacción de la luz con la materia que hace que la luz cambie su energía, la dirección del movimiento o ambas (76)

sonic boom/estampido sónico el sonido explosivo que se escucha cuando la onda de choque de un objeto que se desplaza a una velocidad superior a la de la luz llega a los oídos de una persona (45)

sound quality/calidad del sonido el resultado de la combinación de varios tonos por medio de la interferencia (48)

sound wave/onda de sonido una onda longitudinal que se origina debido a vibraciones y que se desplaza a través de un medio material (31)

standing wave/onda estacionaria un patrón de vibración que simula una onda que está parada (18, 46)

T

translucent/traslúcido término que describe la materia que transmite luz, pero que no transmite una imagen (83)

transmission/transmisión el paso de la luz u otra forma de energía a través de la materia (82)

transparent/transparente término que describe materia que permite el paso de la luz con poca interferencia (83)

transverse wave/onda transversal una onda en la que las partículas del medio se mueven perpendicularmente respecto a la dirección en la que se desplaza la onda (7)

W

wave/onda una perturbación periódica en un sólido, líquido o gas que se transmite a través de un medio en forma de energía (4)

wavelength/longitud de onda la distancia entre cualquier punto de una onda y un punto idéntico en la onda siguiente (11)

wave speed/rapidez de onda la rapidez a la cual viaja una onda a través de un medio (12)

Index

Boldface page numbers refer to illustrative material, such as figures, tables, margin elements, photographs, and illustrations.

A

absorption
  of light, 76, **76, 77, 82**
  of sound waves, 42, **42**
acceleration, 157, 159
air, speed of sound in, 36, **36, 37**
airplane flight, 36
amplification, from lasers, 110
amplifiers, 39, **39, 49**
amplitude, 10, **10**
  interference and, **17**
  loudness and, 39, **39**
  on oscilloscopes, 40–41, **40, 41**
amplitude modulation, 67–68
AM radio waves, 67–68
angle of incidence, 74, **74**
angle of reflection, 74, **74**
anvils, in ears, **33**
apertures, **108**
Archimedes' principle, 158
area, **143,** 156
astronomers, 123
average acceleration, 159
averages, 153
average speed, 159

B

bacteria, 94
balances, triple-beam, 145
bar graphs, 152, **152**
bats, echolocation by, 43, **43**
beluga whales, 42
Bernoulli's principle, 159
bionic eyes, 122
blue sky, 76
body temperature, **144**
bones, **72**
Boyle's law, 158
brain cancer, 122
buoyant force, 158

C

cameras, 108, **108**
cancer, cell phones and, 122
cars
  Doppler effect in, 38, **38**
  mirrors, 99, **99,** 101, **101**
cellos, 46–47, **46, 47,** 49, **49**
cellular phones, 68, 113, **113,** 122
Celsius scale, **144**
Charles's law, 158
charts, 150–152, **150, 151, 152**
chlorophyll, 86
circle graphs, 150, **150**
clapping, **51**
clarinets, 50, **50**
cochlea, **33**
color addition, 85, **85,** 88–89
color deficiency, 106, **106**
colors
  colorblindness, 106, **106**
  mixing light, 85, **85,** 88–89
  mixing pigments, 86–87, **86, 87,** 88–89
  of objects, 83–84, **84**
  in rainbows, 15, **15**
  separation by refraction, 15, 78, **78**
  in televisions, **85**
  visible light and, 70, **70, 71**
color subtraction, 86–87, **87,** 88–89
communication technology, 113, **113**
  cellular phones, 68, 113, **113,** 122
  microwaves and, 68, 113, 122
  radios, 35
  television, **85**
compression, in longitudinal waves, 8, **8, 30, 31**
concave lenses, **102,** 103, **103, 105**
concave mirrors, 100, **100, 101**
concentration, 160
conclusions, 147
cones, 104, 106
conservation of energy, law of, 157
constructive interference, 17, **17**
  in light waves, 80, **80**
  in sound waves, 44–45, **44, 45**
  in standing waves, **18**
control groups, 147
controlled experiments, 147
conversions, unit, **143, 144**
convex lenses, 102, **102, 105,** 116–117
convex mirrors, **100,** 101, **101**
cordless telephones, 113
Crab nebula, **6**
crests, 7, **7, 8**

cross-multiplication, 153
cube volume, 156
cubic centimeters, **143**
cubic meters, **143**

D

data, 151, **151**
deafness, 34
decibels, 40, **40**
decimals, 154
density, 160
dependent variables, 151, **151**
destructive interference, **17,** 18
  in light waves, 80, **80**
  in sound waves, 44, **44**
diffraction, 16, **16,** 79, **79**
diffuse reflection, 75, **75**
digital cameras, 108
dinosaurs, sound from, 58
distance, in light years, **6**
dog whistles, 37
Doppler effect, 38, **38,** 43, **43**
drums, 39, **39**
Dudley, Adam, 59

E

"Ear," 58
ears, **33**
earthquakes, 5, **5**
echoes, 14, 42, **42**
echolocation, 43, **43**
Einstein, Albert, 95
electric fields, 62–63, **62, 63**
electromagnetic spectrum, **66–67,** 66–72
electromagnetic waves (EM waves), 62, **62.** See also light; sound waves; waves
  energy transfer by, 6, **6**
  light as, 62, **62**
  longitudinal, 8, **8, 11, 30,** 31
  producing, 63
  sound waves as, 8, **8**
  transverse, 7, **7, 8, 11**
EM waves (electromagnetic waves), 62, **62.** See also light; sound waves; waves
  energy transfer by, 6, **6**
  light as, 62, **62**
  longitudinal, 8, **8, 11, 30,** 31
  producing, 63
  sound waves as, 8, **8**
  transverse, 7, **7, 8, 11**

Index

Index

Index

Credits

PHOTOGRAPHY

Front Cover James Noble/Corbis; (bkgd), Mehau Kulyk/Science Photo Library/Photo Researchers

Skills Practice Lab Teens Sam Dudgeon/HRW

Connection to Astrology Corbis Images; **Connection to Biology** David M. Phillips/Visuals Unlimited; **Connection to Chemistry** Digital Image copyright © 2005 PhotoDisc; **Connection to Environment** Digital image copyright © 2005 PhotoDisc; **Connection to Geology** Letraset Phototone; **Connection to Language Arts** Digital Image copyright © 2005 PhotoDisc; **Connection to Meteorology** Digital Image copyright © 2005 PhotoDisc; **Connection to Oceanography** © ICONOTEC; **Connection to Physics** Digital image copyright © 2005 PhotoDisc

Table of Contents iv (tl), © Jason Childs/Getty Images; iv (bc), John Langford/HRW; v (tr), ©Digital Vision Ltd.; v (b), ©Cameron Davidson/Getty Images; v–vii, Victoria Smith/HRW; x (bl), Sam Dudgeon/HRW; xi (tl), John Langford/HRW; xi (b), Sam Dudgeon/HRW; xii (tl), Victoria Smith/HRW; xii (bl), Stephanie Morris/HRW; xii (br), Sam Dudgeon/HRW; xiii (tl), Patti Murray/Animals, Animals; xiii (tr), Jana Birchum/HRW; xiii (b), Peter Van Steen/HRW

Chapter One 2–3 (all), © Jason Childs/Getty Images; 5 (tr), Robert Mathena/Fundamental Photographs, New York; 5 (bl), © Albert Copley/Visuals Unlimited; 6 (t), NASA; 13 (tr), © Steve Kaufman/CORBIS; 14 (br), Erich Schrempp/Photo Researchers, Inc.; 15 (tl), Richard Megna/Fundamental Photographs; 16 (tc), Educational Development Center; 18 (tl), Richard Megna/Fundamental Photographs; 18 (bl), John Langford/HRW; 20 (br), James H. Karales/Peter Arnold, Inc.; 21 (b), Sam Dudgeon/HRW; 22 (bl), Richard Megna/Fundamental Photographs; 23 (bl), Martin Bough/Fundamental Photographs; 26 (tl), Pete Saloutos/The Stock Market; 26 (tr), The Granger Collection, New York; 27 (all), Peter Van Steen/HRW

Chapter Two 28–29 (all), © Flip Nicklin/Minden Pictures; 31 (tl), John Langford/HRW; 32 (tr), Sam Dudgeon/HRW; 34 (tr), Sam Dudgeon/HRW; 35 (tr), Mary Kate Denny/PhotoEdit; 36 (all), Archive Photos; 38 (t), John Langford/HRW; 39 (bl), John Langford/HRW; 41 (tr), Charles D. Winters; 43 (t), © Stephen Dalton/Photo Researchers, Inc.; 44 (tl), Matt Meadows/Photo Researchers, Inc.; 46 (all), Richard Megna/Fundamental Photographs; 47 (tr), Sam Dudgeon/HRW; 48 (bc), Sam Dudgeon/HRW; 49 (bl), Digital Image copyright © 2005 EyeWire ; 49 (br), John Langford/HRW; 50 (tr, tl), Digital Image copyright © 2005 EyeWire ; 50 (bc), Bob Daemmrich/HRW; 52 (bl), Richard Megna/Fundamental Photographs; 53 (br), Sam Dudgeon/HRW; 54 (tl), © Flip Nicklin/Minden Pictures; 54 (bc), Sam Dudgeon/HRW; 55 (tc), © Ross Harrison Koty/Getty Images; 55 (cl), Dick Luria/Photo Researchers, Inc.; 55 (br), John Langford/HRW; 59 (all), Victoria Smith/HRW

Chapter Three 60–61 (all), Matt Meadows/Peter Arnold, Inc.; 63 (tl), Charlie Winters/Photo Researchers, Inc.; 63 (tr), Richard Megna/Fundamental Photographs; 64 (t), © A.T. Willett/Getty Images; 65 (tr), © Detlev Van Ravenswaay/Photo Researchers, Inc.; 66 (bc), Sam Dudgeon/HRW; 66 (br, bl), John Langford/HRW; 67 (bcr), Hugh Turvey/Science Photo Library/Photo Researchers, Inc.; 67 (br), Blair Seitz/Photo Researchers, Inc.; 67 (bcl), Leonide Principe/Photo Researchers, Inc.; 67 (tc, tr), Sam Dudgeon/HRW; 69 (br), © Tony Mcconnell/Photo Researchers, Inc.; 69 (tl), © Najlah Feanny/CORBIS SABA; 70 (t), © Cameron Davidson/Getty Images; 71 (cr), © Sinclair Stammers/SPL/Photo Researchers, Inc.; 72 (br), © Michael English/Custom Medical Stock Photo; 73 (tr), Hugh Turvey/Science Photo Library/Photo Researchers, Inc.; 75 (br), © Darwin Dale/Photo Researchers, Inc.; 76 (bl), Sovfoto/Eastfoto; 77 (bl), Richard Megna/Fundamental Photographs; 79 (br), Ken Kay/Fundamental Photographs; 81 (cr), Ken Kay/Fundamental Photographs; 82 (br), Stephanie Morris/HRW; 83 (all), John Langford/HRW; 84 (tl), Image copyright ©1998 PhotoDisc, Inc.; 84 (tr), Renee Lynn/Davis/Lynn Images; 84 (bl), Robert Wolf/HRW; 655 (tl), Leonard Lessin/Peter Arnold, Inc.; 86 (br), Sam Dudgeon/HRW; 657 (t), Index Stock Photography, Inc.; 87 (cr), Peter Van Steen/HRW; 89 (tr), Sam Dudgeon/HRW; 660 (br), Matt Meadows/Peter Arnold, Inc.; 90 (tl), Image copyright © 2005 PhotoDisc, Inc.; 91 (cr), Charles D. Winters/Photo Researchers, Inc.; 91 (bcr), © Mark E. Gibson; 91 (br), Richard Megna/Fundamental Photographs; 94 (tl), Dr. E. R. Degginger; 94 (tr), courtesy of the Raytheon Company; 95 (cr), © Underwood & Underwood/CORBIS

Chapter Four 96–97 (all), Data courtesy Marc Imhoff of NASA GSFC and Christopher Elvidge of NOAA NGDC. Image by Craig Mayhew and Robert Simmon, NASA GSFC.; 98 (b), Yoav Levy/Phototake; 99 (tr), Stephanie Morris/HRW; 99 (bl, br), John Langford/HRW; 100 (tl), John Langford/HRW; 100 (tr), Richard Megna/Fundamental Photographs; 102 (tl, tr), Fundamental Photographs, New York; 106 (tl), © Digital Vision Ltd.; 106 (tr), Courtesy www.vischeck.com (program)/©Digital Vision Ltd. (frogs); 107 (tr), © Yoav Levy/Phototake; 111 (tr), Sam Dudgeon/HRW; 111 (bl), Don Mason/The Stock Market; 112 (all), Victoria Smith/HRW; 113 (cr), © Steve Dunwell/Getty Images; 117 (br), Sam Dudgeon/HRW; 118 (tl), Yoav Levy/Phototake; 118 (br), © Digital Vision Ltd.; 122 (tr), Digital Image copyright © 2005; 122 (tl), M. Spencer Green/AP/Wide World Photos; 123 (bc), Photo courtesy R.R. Jones, Hubble Deep field team, NASA; 123 (cr), NASA

Lab Book/Appendix "LabBook Header", "L", Corbis Images; "a", Letraset Phototone; "b", and "B", HRW; "o", and "k", images ©2006 PhotoDisc/HRW; 124 (br), Sam Dudgeon/HRW; 125 (all), Richard Megna/Fundamental Photographs; 127 (br), Sam Dudgeon/HRW; 128 (br), HRW Photo; 129 (r), Sam Dudgeon/HRW; 131 (br), Sam Dudgeon/HRW; 132 (br), Sam Dudgeon/HRW; 138 (br), Victoria Smith; 139 (br), Victoria Smith; 145 (tr), Peter Van Steen/HRW; 145 (br), Sam Dudgeon/HRW; 159 (tr), Sam Dudgeon/HRW

TEACHER EDITION CREDITS

1E (cr), Erich Schrempp/Photo Researchers, Inc.; 1F (bl), Fundamental Photographs, New York; 1F (cr), Richard Megna/Fundamental Photographs, New York; 27E (cl), John Langford/HRW; 27F (tl), © Stephen Dalton/Photo Researchers, Inc.; 27F (cr), John Langford/HRW; 27F (c), Bob Daemmrich/HRW Photo; 95E (cl), Fundamental Photographs, New York; 95E (bl, cr), © Yoav Levy/Phototake; 95F (cl), Sam Dudgeon/HRW

Answers to Concept Mapping Questions

CHAPTER 1 The Energy of Waves

13.

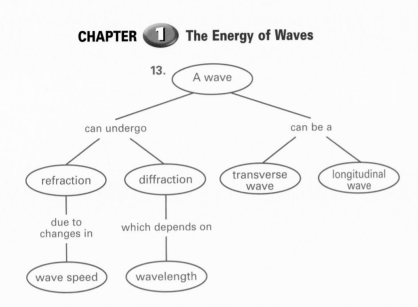

CHAPTER 2 The Nature of Sound

15.

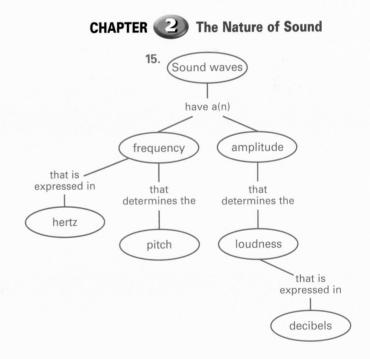

CHAPTER ③ The Nature of Light

18.

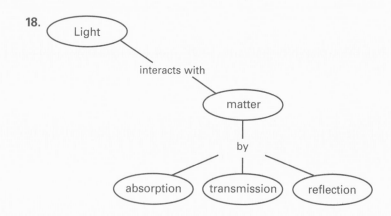

CHAPTER ④ Light and Our World

16.

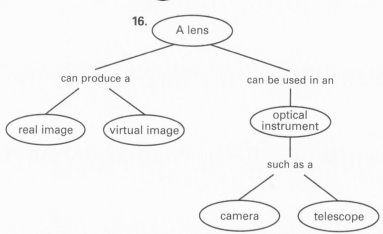